ALTERNATIVES TO PAIN
In Experiments on Animals

Critically ill monkey who has just received electric shock from footplates. (Courtesy: International Primate Protection League).

ALTERNATIVES
TO PAIN
In Experiments
on Animals

Dallas Pratt M. D.

ARGUS ARCHIVES

Cover and title page designed by Dorothea von Elbe

For the Two Mauds

CONTENTS

CONTENTS

Acknowledgments

I am much indebted to Christine Stevens, President of the Animal Welfare Institute, who allowed me to benefit from an extensive literature search on painful experimentation carried out by Jeff Diner in 1977-78. I am also grateful to Dr. Herbert Rackow for his suggestions concerning pain-relieving drugs.

At Argus Archives, Jean Stewart applied her many professional talents to research, proofreading and preparing the lengthy reference list. Eileen Ward typed the copy for the printer with exceptional care.

One

INTRODUCTION

This book is a sequel to another which I wrote earlier in the 1970's and which was published in 1976 under the title *Painful Experiments on Animals*. (Pratt,D.,1976). That volume described numerous biomedical experiments in American laboratories, mostly in New York State. It covered the subject of legal protection of animals (or lack of it), laboratory care, and briefly touched on some of the methods currently available as alternatives to the use of animals.

That publication is out-of-print. Since some of the research it documented is still current, this makes a second appearance in what has now become a selective survey of painful experiments throughout the past decade. But the present volume, unlike its predecessor, describes many experiments from other parts of the United States (not just from New York), and has much more to say about alternatives.

Others have written about alternatives to the use of animals but I think not as they are presented here. The late D.H. Smyth at the suggestion of the Research Defence Society of Great Britain produced in 1978 a book entitled *Alternatives to Animal Experiments*. (Smyth,D.,1978). This is a scholarly but

1

very clearly written summary of many *types* of experiments (although no actual cases are detailed), with an equally lucid description of numerous non-animal alternatives. Anyone who wants a thumbnail sketch of procedures such as chromatography, mass spectrometry, radioimmunoassay, and the use of isotopes will find them simply outlined in Professor Smyth's book. Unfortunately, his advocacy of alternatives, if it can be called that, is so conservative that the book offers little hope or stimulation for the replacement of animals by technology.

A New Approach: Alternatives Matched to Specific Experiments

This writer, on the other hand, is convinced that there are many alternative methods which can be used to reduce the suffering of animals, or to eliminate them entirely from certain experiments. But instead of describing experiments and alternatives in general I have tried a new approach: namely, to describe specific experiments, with excerpts often in the experimenters' own words, and then to match these with specific alternatives whenever I have been able to find them. I hope this will give the reader something more concrete than a vague idea that somewhere the experimenter can find an alternative if he would only look for it. I also hope that it will challenge the scientist to think about what he is doing in causing pain, perhaps unnecessarily, to animals, and to investigate the alternatives suggested. If the suggestions are impracticable, or already out-of-date, I should very much like to hear about it, so that others can be explored. The important thing is to open up a debate, hopefully friendly, between the research community and its informed critics.

To Reduce Animal Suffering

Note that the title of this book is: "Alternatives to *Pain*." If you expect that all alternatives discussed here are designed to eliminate animals from experiments, you may be disappointed to find that some of them are the type which will not replace the animal but which will "merely" reduce the animal's suffering. "Merely" is in quotes in deference to the animal's feelings: I suggest that, to an animal in pain, the use of that word without such qualification would be an affront. It will be a happy day when all animals have vanished from the laboratory, but I don't expect that day to come soon, so rather than take an all-or-nothing approach to animal experimentation, I have included anything which seems likely to help the animal here and now, even if it's "merely" a device (cf.p.127) to facilitate transferring a monkey between cage and restraining chair - but with the ease of transfer helping to shorten the periods of prolonged chair restraint.

"Valuable" versus "Worthless" Experiments: a Criterion for Animal Use?

In any discussion of experiments on animals, someone is likely to inquire, "Isn't the animals' pain justifiable if it prevents equal or greater suffering in humans?" Richard Ryder says you can retort by "demanding to know how many *human* lives are worth laboratory extinction in order to save the lives of a thousand men who would otherwise die of disease." (Ryder,R., 1975). And what about those other moral dilemmas: Is there a justifiable war? Shall we condone capital punishment and harsh prison sentences if they are effective deterrents of brutal crimes? *Et cetera.* The ancient debate of whether the end

justifies the means has never been resolved, so how can any
useful purpose be served by applying it to the problem of animal experimentation? Apes are beginning to communicate with
us through the sign language of the deaf. Dare we ask them
for an opinion?

There are also those who feel that a particular effort
should be made to prevent the use of animals in experiments
related to cosmetics, drug dependence, defense research and
other manifestations of human folly and vice. But if one
accepts that animals should not be involved in what some people have decided are "worthless" experiments, does this not
imply that it may be all right to use them in "valuable" research?

I feel that these subjective attempts to assess experiments on a scale of their supposed value to humans are a poor
substitute for the humanitarian principle that objects to *any*
experiments in which animals suffer. What pain means to an
animal I have tried to imagine in my first chapter: "Aspects
of Pain." What the animals actually are has, for me, been
most perfectly expressed by Henry Beston:

"The animal shall not be measured by man. In a world
older and more complete than ours they move finished and complete, gifted with extensions of the senses we have lost
or never attained, living by voices we shall never hear. They
are not brethren, they are not underlings; they are other
nations, caught with ourselves in the net of life and time,
fellow prisoners of the splendour and travail of the earth."
(Beston,H.,1976,p.25).

We Are All Part of the Problem

If one thinks of animals as Beston did, or with any feeling of respect mingled with compassion, it is easy to become

antagonistic to the experimenters as page after page hits one
with its burden of pain. A reviewer of my earlier book said I
was "angry," and I suppose I was and still am. In a letter to
the psychologist Roger Ulrich, I vented some of this feeling,
adding that I was glad he had discontinued and in a sense re-
pudiated the stressful experiments for which he had become
known (cf.p.43). I was promptly rebuked by Dr. Ulrich: "Do
you take or prescribe drugs?" he wrote. "If you do be damn
careful what you say about anyone else harming animals because
the moment you engage in either of those two behaviors you become a part of the problem of animal painful experimentation."
(Ulrich,R.,1979a).

He is right. We are all guilty, if not as actual perpe-
trators, then as connivers through tacit approval or willful
ignorance of what is going on. I can only say that I write in
an attempt to diminish my ignorance and that of others, to
subject my own and the reader's imagination to the animals'
suffering, to search constantly for alternatives which may
lessen that suffering and, to the extent that we can promote
those alternatives, to help us all become less a "part of the
problem."

Focus of the Book

In this book I have approached animal experiments in
three ways.

First, as painful, sometimes agonizing experiences, which
are going on all around us and which should be described ob-
jectively, if possible in the words of the experimenter. Many
people who consider themselves humane say: "Oh I can't bear to
read about it." I fear that guilt, distaste, sloth and other

emotions underlie the apparent humanity of that statement.
While I realize that these feelings can create formidable resistance, surely there is no substitute for confrontation with
the animals' pain. To accept the animals' sacrifice but to be
willfully blind to the means by which it is procured not only
renders one incapable of helping the animals, but, eventually
must damage one's own self-respect.

Second, as already discussed, I have searched for alternatives and have matched them whenever possible with the experimental procedures. Under the heading "Alternatives" a
complete listing of these will be found in the Index.

Third, I have criticized many procedures not just on humane grounds but because I believe them to be scientifically
defective. These critical comments have been specifically
directed to actual experiments as described by the investigators. I have also cited opinions of scientists who are
themselves critical of certain experiments (cf. reference in
Index to "Scientists' criticism of experiments").

A Major Scientific Flaw: Disregarding the Variables

Certain weaknesses in experiments have been noticed so
often that it becomes apparent they have fairly broad application. For example, despite the attempts to mass-produce
laboratory animals (like mice) so that they are genetically
identical and have exactly the same nurture, most of the animals retain some individuality, and this increases as they are
inevitably exposed to variations in environment. In time, new
staff may take over, animals may fall ill, while other variables: food, temperature, bedding, lighting, cage-mates,

handling, etc., may all skew the interpretation of experimental results. Two factors in particular which are often overlooked are pain and its companion, fear. Inadequate anesthesia and postoperative analgesia, rough handling and the nature of the procedure may all increase the animal's suffering and cause physiological reactions which can further distort the findings (cf.p.14 and Index: "Variables"). Yet rarely do experimenters evaluate these effects. Also, there is the fact that reactions of a mouse differ from those of a cat or human: thus there are interspecies as well as intraspecies variations. But perhaps the most important variable of all is the attitude of the investigator himself.

To illustrate: once the investigator has asked the question which the experiment is designed to answer, he starts hoping, and looking, for evidence which will confirm his hypothesis. The perfectly objective scientist should be equally keen to note evidence which disproves his theory, but so much has been invested intellectually, emotionally, and financially in the experiment by the time it is underway, that very often the experimenter focuses more on the regularities which tend to prove his thesis than on the irregularities (frequently caused by the above-mentioned variables) which might disprove or greatly modify it.

A favorite term of experimenters for laboratory animals, "tools of research," suggests where the trouble lies. For when a scientist picks up a rat in his laboratory, it is not the same as picking up a screwdriver, however much the experimenter may desire to isolate the animal from all that individualizes, "animates" or subtly influences it. If he persists

in this attempt to ignore the animal's context, he becomes
like Henry Beston's "man in civilization," who "surveys the
creature through the glass of his knowledge and sees thereby a
feather magnified and the whole image in distortion." (Beston,
H.,1976,p.25). Roger Ulrich criticized behaviorists who only
see an individual "inside a given envelope of skin" rather
than as "the unique focal point of a network of relations."
(cf.p.46). The investigator is also part of this network of
relations, but he rarely recognizes himself there. John Lilly
says that the latter deludes himself in thinking that it is
possible to be the "objective non-involved observer." (Lilly,
J.,1978,p.87).

Fallacies in Interspecies Comparisons

Scientists who have difficulty visualizing animals in any
context other than the narrow confines of their laboratories
nevertheless don't hesitate to project their findings - "extrapolate" them - to the human condition. Claims that humans
will benefit from the research on animals often appear in rit-
ual fashion at the end of a paper, like a sop thrown to the
funding agency. No doubt the potential benefit to humans is
also stressed in the grant application, witness Aronson and
Cooper's request to the government to fund their sexual experiments on cats as a major contribution to the control of
human "hypersexuality and hyposexuality" (cf.p.56). Unfortunately, the context in which such researchers visualize
human problems and pathology is often as constricted as that
in which they see animals. Or consider the astonishing claim
in the *Yale Alumni Magazine* for Dec. 1976 that the experiments
of Yale psychologists in which cats are electrically shocked

through brain electrodes to attack rats "may help man to
master his own violent instincts!" (cf.p.42). Another scientific fallacy lurks in the suggestion that a liability to cancer in humans may be extrapolated from a known susceptibility
in inbred laboratory rodents, leaving out of account the
diverse "promoter" enzymes which must necessarily start the
process, which vary in different individuals, and which vary
even more between species. (cf.p.84).

You and Me: Wild and Heterogeneous Animals

In fact, the animal experimenter often finds himself in a
Catch 22 situation: the more he recognizes and attempts to reduce the variables in his animal models by developing inbred,
carefully defined strains, and otherwise shields his subjects
from the forces of the environment, the more his material di-
verges from the human who, as K.Z. Morgan suggests, "is a wild
or heterogeneous animal living in many types of environment
with various eating and drug habits, with many diseases and
eccentricities, of various ages and so on." (Morgan,K.,1979,
p.20).

Chapter Outline - and the Index

While I do not know how much animal experimentation is
fruitless because of these difficulties, I think it is apparent that much of it is not only inhumane but scientifically
out-of-date. This volume suggests numerous alternatives: sophisticated technology; microorganisms; human material - from
cell culture to the "whole person"; gene splicing; and many
others. As mentioned earlier, these alternatives will be
found throughout the book matched with specific experimental

procedures. After the opening sections on pain, the experiments are grouped alphabetically in chapters: Behavior - aggression, deprivation and stress experiments; Cancer; Immunology; Inhalation; Primates; Radiation; Surgical - burns, drum
trauma and irritants of the digestive tract; Teratogen Testing; and Testing Biologicals - hormones and vaccines. Then
follow two chapters on the testing of chemical substances, with
a concluding section entitled, "Therapies of Tomorrow."

After the References comes the Index. As already noted,
readers will find the alternatives there, listed under the
above headings and under other categories of experiments mentioned in the text.

Names of organizations have occasionally been abbreviated;
these are identified in the Index.

Two

ASPECTS OF PAIN

1. THE NATURE OF ANIMAL SUFFERING

"Not Can They Reason? nor Can They Talk?
but Can They Suffer?"

The French philosopher Descartes (1596-1650), convinced that animals were mechanistic beings without reason, thought or language, opened the door to centuries of animal torture in the name of science. The 18th century philosopher Jeremy Bentham, whose oft quoted rejoinder heads this chapter, thought that reason and language were irrelevant to the fact of an animal's suffering. Although it is hard to imagine anyone nowadays thinking that a dog's yelp when kicked is merely the "creaking of the machine," there is still an attitude held by some that "animals don't feel pain as we do." In refutation, here is a forceful statement on the subject by two scientists, G.F. Poggio and V.B. Mountcastle (Poggio,C.,1960,p.302).

"We are well aware of the difficulty one faces in labeling as painful any stimulus delivered to an experimental animal, for no introspective report is available. Nevertheless, when a stimulus is by nature destructive of tissue, provokes a defense or escape maneuver accompanied by signs of an appropriate emotional change in the normal working animal, and evokes painful sensations when applied to man, it seems reasonable to regard the stimulus as *painful in nature to both animal and man.* To label such stimuli as 'tissue-destructive, escape-provoking, and emotion-changing', but not to call them painful, seems to us a semantic triviality. There is no

11

a priori reason to suppose that in evolution the perception of
pain appears as a wholly new sensory phenomenon in man."

In fact, such a supposition would be quite illogical in
view of the neurophysiological similarity between humans and
animals. Neurons, synapses - the nerve cells and their con-
nections - and the neuroendocrine mechanisms - the electrical
and biochemical paths of communication - are functionally
identical through many species; it is therefore inconceivable
that there could be radical discontinuity between these patterns of nervous response and the consciousness of pain, mediated in humans by these very mechanisms and of such obvious
value to most if not all animals for self-preservation. The
ways of evolution are often devious and surprising but they
are never unparsimonious.[1]

Misreading the Pain Signals

However, to say that organs have practically identical
functions is not to say that they are identical in structure.
A man and a mouse would both jump away from a hot iron, but
men and mice are put together very differently and are not
often mistaken for one another! Internally, too, although
made out of the same materials, the architecture may be different. This may result in a misreading of the pain signals.
"In animals," says Lumb "the fiber systems subserving pain

[1]
*These comments refer to vertebrates, but invertebrates exhib-
it defensive movements which suggest that they experience
pain. Some, like the octopus, have a highly developed nervous system. The degrees of consciousness among the more
primitive organisms are unknown, but anesthetics have been
developed and should be used at least for all multicellular
forms of macroscopic dimensions. (Kaplan,J.,1969).*

perception are more diffuse than in man and are not so easily
destroyed. Therefore, it should not be assumed that animals
are not perceiving pain because man, in a similar situation,
does not." (Lumb,W.,1973,p.67).

A further difficulty arises because animals may not communicate their distress in a sufficiently 'human' fashion.
Dogs in pain may show much restlessness, but cats may simply
sit in stoical silence.

Still another caution against equating pain-signalling in
man with that in animals springs from the fact that man, a social being, has learned from earliest infancy that a display
of fear and pain to others may bring help. But a similar display in many non-social animals is avoided because it is likely to attract aggression, especially from predators, rather
than assistance. (Baker,J.,1948).

Fear and Pain

The failure to recognize the nature of pain or distress
in animals, and probably the subconscious wish of many investigators to deny that their experiments are stressful, result
in the reluctance to use pain-relievers to the extent that
common humanity would demand. It is not only inhumane, it may
distort research findings. M.R. Chance has shown that variations in the environment of rats, for example, will affect the
results of tests - based on ovary weights - in standardizing
gonadotrophic hormone. These variations included changing the
rats' cages, changing the number of their cage-mates, putting
them with strange rats rather than litter-mates, and so forth.
When such comparatively minor environmental tensions can interfere with test results, it is obvious that unrelieved pain
and fear will cause much wider variations. (Chance,M.,1956).

Harold Hillman, a physiologist at the University of Surrey, England, has enumerated the many physiological reactions of an animal in fear and pain.

"An animal under stress, 1. secretes adrenalin from the suprarenal gland into the blood, which raises the blood pressure and heart rate, mobilizes glucose from muscle and liver glycogen; 2. has violent muscular contractions, breaking down creatine phosphate, and accelerating glycolysis, releasing lactic acid into the circulation; 3. hyperventilates, and thus lowers the blood oxygen tension and increases carbon dioxide; 4. increases its metabolic rate and temperature, and changes the pattern of a series of enzyme reactions, each of them dependent to different degrees on the temperature. In summary, there is hardly a single organ or biochemical system in the body which is unaffected by stress....The changes in some of the parameters cited may be up to 300% of the normal values at rest. This would obviously invalidate painful experiments claiming to examine any changes less than those already known to be due to stress itself. It is almost certainly the main reason for the wide variation reported among animals upon whom painful experiments have been done." (Hillman,H.,1970).

An animal under stress shares all the above reactions with man. The actual perception of pain in man and presumably in animals is influenced by what H.K. Beecher has called the "reaction component," the psychological evaluation of the stress and its significance to the sufferer. Soldiers wounded in combat who interpret the trauma as a fortunate means of release from dangerous service often feel only minimal pain. Volunteers in painful experiments, since no serious outcome is anticipated, also, like the soldiers, suffer relatively little. Analgesics, since they act primarily on psychological distress, are not very effective in these cases.

But when the reaction component is fraught with fear and anxiety as in the case of "pathological pain" from conditions such as cancer - pain of serious and depressive significance -

then suffering is much greater, although, fortunately, narcotics are correspondingly more effective. In animals *all* pain is serious and significant, like pathological pain in man. Therefore analgesics are especially needed in animals and are very effective. (Keele,C.,1962).

2. PAINFUL ELECTRICAL STIMULATION OF THE BRAIN

The term "vivisection," with its etymology derived from "alive" and "to cut," originated during an historical period when experiments were chiefly operations on unanesthetized animals. Although those who call themselves "anti-vivisectionists" today define it much more broadly, the term has greatly obscured the fact that the vast majority of modern pain-inflicting experiments use techniques unknown to our forbears. An example is the electrical probing and shocking ("stimulating" to use the more seemly language of the investigator) of areas, nerve cells or fibres in the brain which register pain. Antiquated laws to prevent "surgical" pain frequently result in the anomaly of an animal having to be carefully anesthetized for the implanation of electrodes through scalp and brain coverings, only to be brought out of anesthesia to experience the unimaginable suffering of direct electrical stimulation of the pain centers in the brain itself. For instance, consider the painful experiments performed on ten cats at the State University of Iowa, Iowa City, in 1976.

"These studies," the university reported, "involve the tracing of central pain pathways in chronically implanted electrodes in the brain of the cat. Tranquilizers or

anesthetics would negate the pain and thus would prevent the
fulfillment of the protocol of the research. The pain is produced only during 60 second or less time periods when the animals are either stimulated through chronically implanted tooth
pulp or intracephalic electrodes." (USDA/APHIS,1976,Iowa,Univ.).

) I wrote "unimaginable suffering", above, because humans
are never subjected to such experiments (even specialists in
political torture have not yet introduced the technique, although no doubt they will).[1] Furthermore, pain suffered by an
animal is always unmitigated by the 'philosophy' whereby man
can often temper his perception of pain.

An experiment of this type by R.S. Kestenbaum of the
State University of New York at Stony Brook, E.E. Coons of New
York University and another investigator, used 19 rats, the
pain receptors in whose brains were electrically stimulated by
surgically implanted electrodes. To be sure that the implantation had reached a pain center, they only used those animals
which showed "clearly aversive response to the stimulation,
such as squealing, defecation, urination and running." Each
"one-minute train of noxious midbrain stimulation" could be
interrupted, but only for 3 seconds, if the rat pressed an escape lever. The investigators reported that "due to the
stressful nature of the stimulation some subjects could only
be tested for brief sessions. Thus, an experimental session
consisted of 20-60 1-min. brain stimulation trials, each followed by a 1-min. rest." It is not stated just how "some
subjects" made it known that the agony, prolonged sometimes

---

[1]*Nashold et al. probed pain centers in the brains of conscious
patients, but only as long as the latter would endure it.
(cf.p.19-20).*

up to an hour, was more than they could bear. In such an experiment there was no question, of course, of using a pain-
relieving drug. (Kestenbaum,R.,1973).

The Psychophysiology of Pain

The description of the above experiments brings us into
the laboratory, and in the remainder of this chapter I shall
examine a few of the procedures which are used in brain experiments to investigate the nature of pain. I have mentioned some of the psychological determinants of the pain
experience in humans and what these are, presumably, in animals. But since many experimenters are preoccupied with
neurophysiological and neuroanatomical explorations of the
brain and central nervous system, I shall broaden the discussion and attempt to attach physiology to psychology as an
introduction to these pain experiments.

If our skin is burned or traumatized two things happen:
we are conscious of a very unpleasant sensation and we jerk or
jump away from the source of the pain. We may quickly rub the
skin, or run cold water on it to diminish the pain. If at the
time of the hurt we had been under some strong emotion, or if
our attention had be intensely fixed on something else, we
might at first not have noticed the pain very much.

These reactions can be explained very roughly by the
following: although the sensation of pain and the muscular
reaction appear to follow the trauma almost instantaneously,
in fact a good many things happen between the moment of trauma and the response. The feeling of pain, its intensity, can
be decreased - and it can also be increased - by messages
which rush down from the brain, messages moulded by the

factors of memory, attention and emotion. (The messages con-
veying emotion, motivating "flight or fight," mostly come from
the "limbic" system in the midbrain; attention and memory messages come chiefly from the higher cortical areas.) Hypnotic
analgesia, for instance, is caused by the mind's attention
being totally absorbed, through hypnotic suggestion, in something other than the sensation of pain. An impulse to flinch
from a hypodermic can be inhibited by a memory that the anticipated pain will not be so bad after all. And the sensations
from rubbing or chilling the skin can inhibit the pain, if the
latter is not too intense, by moving much faster to the brain
through the large-diameter fibers which convey them than does
the pain sensation through its small fibers. Thus the brain
can be triggered into inattention and the pain does not register. Acupuncture probably works via the same mechanism:
pricking by needles travels as nerve impulses in large fibers
and is felt in the skin as a discrete stimulus which is pre-
dominant over diffuse pain - again, if the latter is not too
intense.

The neuroanatomical structures forming the substrate of
these responses have been examined by researchers over many
decades in both humans and animals. For example, K.V. Anderson and associates at Emory University, Atlanta, gave young
cats electric shocks to the paws to activate the discrete pain
pathways, and shocks to the upper canine teeth to activate the
diffuse pain pathways. The cats were trained to jump over a
small barrier to escape foot shock, but when the shocks to the
teeth were given at the same time, the "animals assumed a
fixed posture with limbs extended until both stimuli were

terminated." In continued trials pairing tooth and foot shock,
escape responses were absent, even though foot shock was
raised to 14mA - "the maximum amount of current that could be
generated by our apparatus." The experimenters convinced
themselves that 1 mA shocks to the teeth did not hurt the cats
too much, but apparently they hurt enough to reverse the acupuncture effect - the predominance of discrete over diffuse
pain stimuli. (Anderson,K.,1976).

Painful Brain Stimulation in Humans

Memory was mentioned above as one of the contributions
which the brain makes to the shaping of incoming pain sensations. When a limb has been amputated, sometimes the infusion
of memory appears in the shape of "phantom limb", which gives
the patient the sensation that the amputated extremity is still
present. Since this phenomenon is usually accompanied by severe pain, neurosurgeons attempt to relieve it by cutting one
of the nerve tracts which formerly connected the brain and
the limb. B.S. Nashold, Jr. and associates at Duke University
Medical Center reported on several cases of phantom limb which
had been unsuccessfully treated by this type of surgery, as
well as others who were also suffering from intractable pain
of central nervous system origin. All of these Nashold *et al.*
planned to treat by interrupting the ascending pain tracts,
but only after electrode exploration in the conscious patient
of the exact areas involved in the pain sensations. Their report of these preliminary explorations has given us an in-
valuable insight into the character and degree of pain produced by electrical probing of the pain centers in the midbrain or "mesencephalon." The results can be summarized

as follows: 1. stimulation of the gray matter immediately sur-
rounding the central canal in the midbrain ("periaqueductal
gray matter") produced diffuse pain with feelings described as
"fearful", "frightful" or "terrible"-followed by refusal by
the patient to allow it to continue. 2. Stimulation of the
areas lateral to this caused a "brighter" or "sharper" sensation of pain, localized on the opposite side of the body,
with the patient allowing repeated stimulation because the
pain although intense was bearable. (Nashold,B.,1969).

We now know why the monkeys stimulated at Yale University
by José Delgado in the same central midbrain area reacted with
"high pitched vocalization accompanied by grimacing, restlessness, and attempts to grab and bite objects within reach...
with signs of aggressiveness directed against the investigator." (Delgado,J.,1966).

*Risks and Precautions in Experiments on Animals
Paralyzed by Curariform Drugs*

An Editorial Note entitled "Use of Curariform Agents" in
the journal *Experimental Neurology* (Doty,R.,1975), by R.W.
Doty, contains the following admonitions, with a reference (4)
to the Nashold article:

"At one time it was common practice to apply electrical
stimulation to the brain stem of curarized animals...with ro-
ving electrodes. A large body of evidence has now accumulated
on experimental animals and human patients (4) showing that
even moderate electrical stimulation of certain regions of
the limbic system or mesencephalon can produce aversive effects of extraordinary intensity. While it is not common to
strike such loci when using relatively weak stimuli, there is
also no certain way of avoiding them, especially in the mesencephalon. The relative inaccuracy of stereotaxic placements
and the need to use stimuli likely to be suprathreshold for
the "aversive areas" should they be encountered, makes the

risk significant that, sooner or later, such regions will be
stimulated. In addition, there is probably no reliable method
for knowing in the curarized animal whether the stimulation is
aversive. Aware of these facts, few experimenters choose to
move stimulating electrodes through the brain stem of conscious, curarized animals, any more than they would perform
major surgery under these conditions."

Earlier in the editorial, Doty described how an animal's
head, in these brain experiments, can be precisely positioned
by metal clamps (stereotaxis), but warned that a paralyzing
curariform drug will abolish the signs - dilatation of the
pupils, response to pinprick stimuli, etc. - by which the presence of pain is indicated. He adds:

"The widespread success with extracellular or even intracellular recordings from single units in the brain of behaving
animals clearly shows curarization is, in certain instances,
no longer needed for these delicate and difficult procedures
per se."

For just such an experiment, see description of E.V.
Evarts' work on p.128,130,131.

The second reference (2) in this editorial directs the
reader's attention to an experiment on squirrel monkeys which
J.H. Bartlett and the writer of the editorial, Dr. Doty, performed at the Center for Brain Research of the University of
Rochester, New York. Describing the technique actually used
in the experiment, he says:

"Where it is necessary to stimulate the brain stem of the
conscious, immobilized animal, the electrodes should first be
implanted and subsequently the alert, freely moving animals
should be carefully observed for any evidence of aversive effects, e.g., hypermotility, when stimulation is applied to
these electrodes at the intensity to be used in the final experiment (2). With this precaution of stimulating through
fixed electrodes with parameters demonstrated by the animal's
behavior to be innocuous, the possibility of applying aversive
stimuli to an immobilized preparation is precluded."

In spite of his editorial reservations about the use of curariform drugs, Doty apparently felt the monkeys in this experiment had to be paralyzed. Presumably this was to immobilize the eye and eliminate variables in the brain cells' response to light stimuli. The object of the experiment was to study the way electric stimulation of another part of the brain (the mesencephalon) could alter the response of the light-sensitive cells. Although the electrodes penetrated an area whose stimulation in man the investigators admit is "excruciatingly painful," they maintain that the precautions described above protected the animals from discomfort. (Bartlett, J.,1974).

The admission by a neurophysiologist of Doty's stature, confirmed by the editorial in *Experimental Neurology*, that excruciating suffering may be caused by these techniques unless careful precautions are taken, is certainly disturbing. J. Diner abstracts a series of experiments on animals paralyzed by curariform agents. (Diner,J.,1979,p.45-54). Typically, under general anesthesia, the animal's head is fixed to a stereotaxic instrument, tubes are passed into the windpipe and veins, the skull is opened and an electrode is introduced into the brain, and all wound edges are infiltrated with a local anesthetic. Then the general anesthetic is withdrawn and the animal is paralyzed with the curariform agent. The electrode produces recordings and/or stimulation, sometimes of single nerve cells. Often it is moved around, and it is here that the danger of hitting a pain center occurs.

The precautions described by Bartlett and Doty are not often mentioned by other investigators. Therefore all journals that print reports of experiments using curariform agents should adopt the practice of *Experimental Neurology,* which states editorially that the following constitutes a requi.re-ment for consideration for publication:

"For any experiment in which curariform agents are em-ployed, the necessity for their use must be justified and details provided as to steps taken to reduce or avoid distress to the animal, particularly with regard to electrical stimula-tion. (Doty,R.,1975,p.iv).

3. DRUGS WHICH RELIEVE PAIN

Since this book is about pain and how it can be relieved, the subject of drugs as pain-relievers naturally comes up for discussion. Below are brief descriptions of the principal drugs and their actions.

Analgesics are drugs used for the relief of pain without causing unconsciousness or sleep - e.g. aspirin; morphine (a narcotic analgesic).

Hypnotics or *sedatives* are drugs which depress the cen-tral nervous system; they do not relieve pain, they dull the conscious perception of it, tending to produce sleep or un-consciousness - e.g. barbiturates.

Tranquilizers alleviate agitation or anxiety, promote muscle relaxation, make pain more bearable, but do not dull mental acuity - e.g. meprobamate ("Miltown"); chlordiazepox-ide hydrochloride ("Librium").

Anesthetics are drugs which combine any or all of the above actions, leading to the complete elimination of pain and the induction of unconsciousness - e.g. ether; halothane; bar-biturates in large ("anesthetic") doses, or in repeated,

smaller doses. The first two are inhalant gases, particularly
suitable for long operations since the anesthetic level can be
kept constant by altering the inspired concentration; barbiturates are injected, so-called "fixed agents." "Morbidity and
mortality in animals are much greater following the fixed
rather than the inhalation agents, probably due to the long
recovery time, heat loss, pneumonia, etc., aggravated by lack
of postoperative care." (Barnes,C.,1973).

Curariform agents are drugs which produce generalized
muscular relaxation or paralysis but which have no effect on
pain sensibility - e.g. curare; succinylcholine; gallamine
triethiodide ("Flaxedil"). When they are appropriately combined with an anesthetic, muscle relaxation is obtained which
is desirable in surgical procedures, but there is a risk of
causing extreme suffering with curariform drugs, as described
on p.20ff., if analgesia is inadequate or absent. That there
is real cause for concern is illustrated by the finding that
of 110 applicants for grants proposing to use curare in animal experiments, only 7% indicated that it would be accompanied by an anesthetic. These were approved and funded applications made in 1972 and 1976 to either the National Science
Foundation or to one of the institutes of the Alcohol, Drug
Abuse and Mental Health Administration. (Fox,M.,1979,p.11).

Postoperative Analgesia

In general, the various stages of recovery from an operation proceed in this fashion. The immediate postoperative
period, even with the return of consciousness, is usually not
pain-filled. Later, if discomfort begins, light analgesia
may be tried. Still later, if pain increases, perhaps

betrayed after chest surgery by an animal's shallow respiration in an attempt to "splint" the thorax, a stronger narcotic is called for. By this time any respiratory depression
caused by drugs used during the operation itself should have
worn off, thus reducing the chance of pneumonia. (Rackow,H.,
1979).

Barbiturates, often used for animal anesthesia, need special attention, because, as they are metabolized, excreted and
as they decline to less than anesthetic levels, they actually
increase pain sensitivity. If no postoperative analgesic has
been given, an animal may experience considerable pain as the
barbiturate concentration in the brain diminishes. The same
is true for halothane. (Croft,P.,1964).

Withholding Postoperative Analgesics

During 1973, postoperative analgesics were rarely if
ever ordered at New York State Veterinary College at Cornell.
(Rubin,N.,1974). No doubt it was the same in other years.
Animals there are used for multiple procedures, so the repeated experience of recovery without pain-relief adds up to considerable suffering. In 1966, testifying before a subcommittee of the House Committee on Agriculture, Ralph Mayer stated
that during his employment as a technician at the Minneapolis
Veterans Administration Hospital no postoperative pain-
relievers had "ever been given...to any dog, including the
major surgery cases, such as gastrectomies, lung transplants,
kidney transplants, bowel anastomoses, open-heart surgery and
brain surgery." (US Congress,House of Reps.,1966).

At the same House Hearing, Dr. Donald McQuarrie, Director
of Experimental Surgery at the Minneapolis hospital, attempted

to defend the withholding of analgesics on the grounds that
"narcotics in a dog depress the respiration...so that pneumonia
occurs very frequently, and death is common."

Dr. McQuarrie's comment is surely relative to special circumstances at his hospital. In what condition were the dogs
preoperatively? Were they random-source animals, obtained by
the hospital from pounds, perhaps already suffering from respiratory disease or anemic from worms? Were his animal-care
facilties understaffed, so that postoperative observation was
not available to check on the dogs' condition? Were "fixed"
rather than inhalation anesthetics preferred by the surgeons,
with their greater risk of postoperative complications? Were
analgesics given too soon after surgery, before recovery from
the anesthetic was well advanced? Was there good rapport with
the animals before the operation? Barnes and Eltherington's
text comments on this:

"'Making friends' with the animal will contribute much to
a smooth anesthetic induction. Attempting to anesthetize a
frightened, struggling animal, often made that way by undue restraint, may lead to fatal consequences due to the use of more
anesthetic than normally required." (Barnes,C.,1973,p.15).

Although it is probable, for a variety of reasons ranging
from the matter of expense to sheer inhumanity, that postoperative analgesics are often withheld from animals when they
would have been prescribed for human patients, it would nevertheless be unreasonable to insist on their use in every case.
For instance, it has been found that, in a consecutive series
of over 1000 human patients who had had general surgical or
urological operations, 36% had no need at all for any analgesic drug in the entire postoperative period. (Jaggard,R.,
1950). Furthermore, the proportion of patients requiring

analgesia postoperatively correlates with the site of the operation, as follows: chest cavity, 74%; upper abdominal, 63%;
lower abdominal (excluding gynecological), 51%; limb operations, 27%; inguinal, 23%; body wall, 20%; neck, 12%. (Loan,
W.,1967).

Other factors also influence the severity of postoperative pain. They include fear, nausea, drainage tubes, abdominal distension and adequacy of dressings.

These matters must all be weighed by those ordering pain-
relieving drugs. Each case is different and there is no simple solution to relief of suffering in animals under experiment. In the operative and postoperative situation there are
some fairly precise do's and don'ts, as the above outline
suggests. There are less precise indications in nonsurgical
cases: for instance, in attempting to relieve the chronic
pain of cancer. And unfortunately there are many experiments
such as pain, stress, and behavior studies; toxicity tests;
safety and potency assays of drugs; addiction studies and investigations of many diseases; in which pain-relieving drugs
are withheld on the grounds, sometimes valid but often merely
traditional or statutory, that giving them would "defeat the
purpose of the experiment." Comments on this will be found
throughout the book.

4. THE MOST PAINFUL, DISTRESSING, FEAR-INVOKING
EXPERIMENTS

In this book many types of painful experiments are described, some at greater length than others. It is in fact a
survey of such experiments throughout the 1970's. For the

sake of at least relative completeness, a listing follows of
procedures or induced maladies which have been singled out by
various authorities as particularly painful, distressing or
fear-invoking.

The sources drawn upon variously described these procedures as ones causing "pain or distress," or "gave concern,"
or in some cases were "not acceptable," especially if carried
out too rigorously or for too long a time or without appropriate pain-relieving drugs. The experiments or illnesses cited
inflict a degree of suffering on animals which, depending on
the circumstances, ranges all the way from "severe" to the extremity of torture. The authorities include a group of research facilities listing experiments - in their Annual Reports
for 1976 under the Animal Welfare Act - which caused pain or
distress to animals without the benefit of pain-relieving
drugs; a spokesman for the Inspectorate of the British Home
Office; Dr. J.R. Baker and a group of British scientists; the
Universities Federation for Animal Welfare; the Animal Welfare
Foundation of Canada and the Canadian Council on Animal Care;
the National Society for Medical Research; the Animal Welfare
Institute, as sponsor of *Physical and Mental Suffering of Ex-
permental Animals* (Diner,J.,1979), and Dr. Alice Heim, recently Chairman of the Psychology Section of the British Association for the Advancement of Science. These groups represent
either research-associated scientific bodies or animal welfare organizations identified with a reformist rather than an
anti-vivisectionist position; thus their selection of experiments as ones which often lead to undue suffering is particularly significant.

The experiments are listed alphabetically; the reference numbers refer to the above-mentioned authorities (cf. key on p.30); the letters "AP" identify citations in this book (cf. Index).

Anaphylactic shock.
4; AP.

Aggression, induced (often by electric shock - resulting in fighting, wounding, killing).
2, 9; AP.

Blinding.
9; AP.

Brain stimulation, electrical, painful.
6, 7, 9; AP.

Burns.
2; AP.

Cancer, chronic.
1, 2, 7; AP.

Cannula, chronically implanted.
5; AP.

Cold, prolonged exposure to.
2, 7; AP.

Coronary constriction, induced.
9; AP.

Defense research (radiation, irritants, fire, explosives, etc.).
7; AP.

Drowning, following swimming to exhaustion.
7, 8; AP.

Drug addiction (withdrawal symptoms).
8.

Electric shock, painful.
6, 9; AP.

Fear ("conditioned emotional reaction.").
8, 9; AP.

Fractures.
2.

Grooming disruption.
9; AP.

Heat, prolonged exposure to.
2.

Heart failure, stress induced.
9; AP.

Helplessness, learned (usually via electric shock).
8; AP.

Infectious disease, chronic.
2, 7; AP.

Inhalants, toxic.
5; AP.

Irritants, corrosive, of digestive tract.
4, 9; AP.

Irritants, eye.
1, 4, 5, 9; AP.

Irritants, joint (injected).
4, 5; AP.

Irritants, skin.
4, 5; AP.

Isolation, social, or from mother.
8, 9; AP.

Pain, postoperative, un-
relieved.
2, 7; AP.

Pain research.
9; AP.

Pain, unrelieved, combined
with curariform paralysis.
2, 7, 9; AP.

Pathogens (Rabies, Tetanus,
etc.), virulence challenge.
4; AP.

Punishment (usually electric
shock) in behavior con-
ditioning.
5; AP.

Radiation sickness.
1, 9; AP.

Repeated use of same animal (especially in painful brain
exploration).
7; AP.

Restraint, physical, prolonged.
2, 7; AP.

Seizures, convulsive, induced by
chemical or electrical brain
stimulation.
9.

Sleep deprivation, prolonged.
8.

Sound, loud, prolonged exposure
to.
2, 9.

Starvation, prolonged.
2; AP.

Stress.
6, 9; AP.

Thirst, prolonged.
2; AP.

Tooth-pulp stimulation, painful.
4, 9; AP.

Toxicity Testing, in general.
2, 4, 5, 7; AP.

Toxicity Testing, lethal dose
("LD/50").
5; AP.

Toxicity Testing, metal poisoning.
7; AP.

Toxicity Testing, rodenti-
cides.
6.

Trauma (battering in drum;
crushing, striking) without
pain relief.
2, 3, 9; AP.

Ulcers, gastric, stress induced.
8; AP.

REFERENCES

1. U.K. Home Office, 1974.

2. Canadian Council on Ani-
    mal Care, 1978.

3. Baker, J., 1949.

4. USDA/APHIS, 1976.

5. Anon., 1975.

6. Universities Federation
    for Animal Welfare,
    1963.

7. Hughes, T., 1976.

8. Heim, A., 1978.

9. Diner, J., 1979.

Three

BEHAVIOR EXPERIMENTS

1. INTRODUCTION

The method of studying the behavior of animals used by many investigators is, paradoxically, not to study animal behavior. What they seem to be interested in are the distortions, the pathology of behavior: either the fragments which remain after surgical or other mutilations have destroyed the marvelous wholeness of a functioning organism, or the reflex jerks teased out by any of the myriad of prods, punishments, or pleasures which the ingenuity of a researcher can devise. The wholeness which the ethologist prizes in the wild is useless to the investigator in the laboratory. Perhaps his mind is affronted by the freedom of the wild animal, by the strength the latter draws from the earth beneath its feet and in the labyrinth of its natural habitat. He is uncomfortable until the animal is subdued, and transformed.

The instruments of transformation are the drug, the knife and the electrode. Add to these the cage, with its accessories: the shuttle box, response levers, pellet dispensers and stimulus panel; and the restraint system: stereotaxic head holder, primate chair, *et cetera*. Thus equipped, the behaviorist is ready to proceed with his experiment. If he deploys much technology and uses it ingeniously enough his colleagues may described his work as "elegant," the scientist's adjectival accolade.

The experiments described in this chapter are of three kinds. The first group are studies of aggression provoked in the animal by fear or pain - reactions usually produced by electric shocks. The aggression is allowed an outlet, either through attacking another animal, or biting whatever comes to hand.

The second group are deprivation experiments, which physically and often surgically deprive the animal of one of its senses, or separate it from some part of the environment necessary for its survival.

The third group are stress experiments, in which the animal is again exposed to fear and pain, but is frustrated in its attempts to escape, resulting in psychological, and sometimes physical, disintegration.

2. ARTIFICIALLY STIMULATED AGGRESSION

It will be apparent in the studies of animal aggression which are reviewed below that the behavior described is generally elicited by human manipulation provoking pain, fear or rage.

Prey-killing and Fighting

There is a large literature describing experiments promoting mouse killing by rats ("muricide"), or frog killing ("ranacide"), or mice fighting mice, or rats fighting rats, as a means of studying aggression.

This behavior has been explored in numerous ways. Electric shocks are delivered to the feet of the animals or directly into the brain through implanted electrodes, aggression-stimulating drugs are injected, parts of the brain, such

as the olfactory bulb, are surgically removed, and rodents
from the outside are introduced into a resident colony. After
a certain time, the dead animals are counted and the distribution of wounds is noted. Finally, the animals who have been
forced to become aggressors are killed so that the changes if
any in their brains can be evaluated.

These fights between animals (which frequently show no
natural antagonism) are stimulated by a group of investigators
who, if they were promoting fights between larger animals, such
as cocks and dogs, would be prosecuted under the Animal Welfare
Act. But Congress has specifically denied these small animals
the protection of the law. As a result, ingenious variations
of these laboratory massacres, and portentous sounding papers
(e.g. "Role of the mystacial vibrissae in the control of isolation induced aggression in the mouse.") (Katz,R.,1976) earn
the researchers masters' degrees, doctorates, and never-fail-
ing largesse from the National Science Foundation, National
Institute of Mental Health and other tax-supported agencies.
Katz's work on "mystacial vibrissae" (or as the uninformed
would call them, mouse whiskers), was, for example, funded by
the National Institute of Mental Health through the Mental
Health Research Institute, University of Michigan.

Since this and similar experiments are often partly supported by public funds, members of the public, *if they knew
what was going on,* might well inquire just how such work "will
undoubtedly contribute to the benefit of Man or animals" - to
use the criterion which the Canadian Council on Animal Care
applies to prey-killing and fighting experiments (singled out
as ethically problematical because of the suffering involved).

Experimenters who have built their careers on promoting
fights between rats and rats, rats and mice, rats and frogs,

cats and mice, etc. may feel aggrieved to have their efforts
criticized, yet without compunction they expose sensitive, unanesthetized animals to degrees of suffering which are comparable to those of a medieval torture chamber. The following
examples are chosen from many which the reader may peruse in
the original; some are abstracted in Jeff Diner's *Physical
and Mental Suffering of Experimental Animals,* a review of the
pertinent scientific literature. (Diner,J.,1979).

Attacked for 21 Hours; Wounded 68 Times (Average)

In 1977, psychologists from Rutgers University, New Jersey, and Ithaca College, New York, reported an experiment in
which stranger male rats were placed singly in a cage containing an established colony of male and female rats, some of the
latter with sucklings. The intruders were attacked by the res-
idents and each received "an average of 50.9 small (less than
1/2cm.), 12.6 medium (0.5 to 1 cm.), and 4.6 large (more than
1 cm.) body wounds." After 21 hours of this torment, the man-
gled intruders were killed and their stomachs and small intestines were examined. The findings were significant, because,
for those inclined to discount a rat's sensitivity to pain, the
autopsy report of "ulcers and mucosal degeneration of the small
intestine;...erythema [inflammation], gastritis and prominent
rugae [folds] in the stomach" is objective evidence of the extreme stress and suffering undergone by the victims. This is
emphasized by the development of so much pathology after less
than a day's exposure to punishment. (Lore,R.,1977).

The investigators, probably becoming aware of the cold
draught of criticism of these endless studies of aggression in
laboratory rats - the journals have been full of them for forty
years - express the hope that their "discovery of a positive

relationship between attack stress and gastrointestinal functioning will stimulate the use of animal models with ecologi-
cal validity for the study of stress and its physiologic con-
comitants." As Diner drily comments, "It is rather difficult
to imagine an animal in a natural setting subjecting itself to
21 hours of continuous stress." (Diner,J.,1979,p.163). However, the Blanchards, a husband and wife team at the University
of Hawaii, who have been generating much research and many
papers in recent years on electric shock-stimulated rat fighting, have reassuring words for colleagues who might be alarmed
at the thought of abandoning their long-suffering domesticated
animal models for anything more natural:

> "One very pleasing aspect of the specific agonistic be-
> haviors indicated by these studies of laboratory rats is their
> similarity to descriptions of wild rat agonistic behavior....
> This finding therefore provides considerable justification for
> the continued use of the laboratory rat as a research animal
> in the study of attack and defense....Calls continue to be made
> for the abandonment of domesticated rats in favor of more 'nat-
> ural' subjects, a tactic which the present findings suggest may
> be unnecessary." (Blanchard,R.,1977).

Alternatives in a Natural Laboratory

Pleas not only to study "wild" animals, but to observe
them in their natural setting, have been heard for a long time
and now are reaching an increasingly sympathetic and concerned
public. Twenty years ago Russell and Burch cited many experiments in the literature of ethology in which fear, occurring
in the *normal* life of animals (e.g. a pigeon's fear of aggression when it finds itself in another's territory), might be
studied humanely. (Russell,W.,1959,p.145-153).

In recent years, the work of ethologists like Jane Goodall
and George Schaller, based on observations in the wild, has
been very informative, and includes much material on prey-
killing and intra- and inter-species fighting. Goodall's book

Innocent Killers (written in collaboration with her former
photographer-husband, Hugo van Lawick) is a masterly study of
wild dogs, jackals and hyenas. (van Lawick,H.,1970,p.209).
After two years' observation in Ngorongoro and Serengeti, Tan-
zania, the animals the authors followed so patiently had all
acquired recognizable personalities and, indeed, names. Compare the description of the subjects of a laboratory rat experiment: "six male albino Sprague-Dawley rats, 150 days old
and purchased from Holtzman Co." with a retrospective paragraph about the jackals studied in *Innocent Killers*:

> "The memories of that period are, for us, many and vivid:
> the boisterous tumbling play of Rufus and Nugget and their
> sister Amba; the way in which Cinda, the runt, so frequently
> curled up by herself; Cinda in the talons of the eagle, and
> her shrill screaming as she hurtled to the ground; Jewel,
> their mother, pouncing on cub after cub, bowling it over, and
> then grooming it until finally it escaped to join the games of
> its siblings; Jason darting in and out of lion and hyena kills
> for titbits; Jason battling with a snake; Jason, alone, chal-
> lenging and driving off a huge lappet-faced vulture from the
> food of one of his cubs." (*Ibid.*,p.209).

The Tanzanian lions in Schaller's *Golden Shadows, Flying
Hooves* (Schaller,G.,1973), also acquired recognizable personalities. Again, the book is full of insights into the aggressive
and predatory habits of these animals. As one reads the reports of laboratory observations of prey-killing and fighting
among rats, one is struck by the extreme narrowness of viewpoint - it does indeed seem to be a case of learning more and
more about less and less. In the wild, as Schaller points out,

> "No creature stands alone; the scope of the study soon
> broadened to include not only the lion's predatory associates
> such as the hunting dog, but also the array of prey species on
> which the carnivores depend for survival."

The wide open spaces of Nature's laboratory, where the unexpected is always welcome and instructive, contrast with man's

claustrophobic work space, where the unexpected is feared because it may confuse the experiment's simplistic design.

Effect on Aggression of Electric
Shock (in Dirty Cages)

Unfortunately, we must leave the fresh air of the African
plains and return to a particularly claustrophobic center of
aggression research: the Psychology Department of the University of Iowa. There, a group of investigators have developed
a plexiglass "fighting chamber" in which a rat selected as the
target of aggression is restrained in a leather and metal har-
ness and exposed to another rat stimulated to attack him by
the receipt of as many as 100 electric shocks, up to an intensity of 2.5 mA. The target rats are also shocked (by electrodes attached to their hind paws) to provoke a display of
threat behavior.

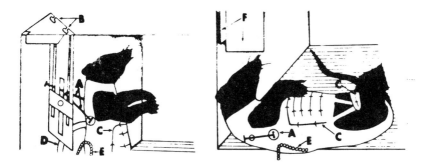

Fig. 1. Left: harnessed rat restrained upright; right: su-
         pine. Harness grids (C) are wired to electrified
         grid floor by cable (D). Metal plates (F) deliver
         shocks to aggressor's forepaws if aggressor stands
         on target rat's head. (Copyright,1976,Society for
         the Experimental Analysis of Behavior,Inc.).

In a first experiment, 18 rats were used, in the second,
no less than 408. The conclusion was that the more the restrained rats were goaded by increasing intense shocks into
threat display, the less they were attacked by the others.
However, the rats restrained in a supine rather than an upright position were apparently seen as less threatening and
were therefore more frequently attacked, in some cases up to
200 times during three five-minute sessions. (Hynan,M.,1976).

Electric shocks and exposure to attacks by other rats
were not the only source of suffering of animals in this institution. The Dept. of Agriculture's Animal and Plant
Health Inspection Service reported severe sanitation problems
there in Dec. 1978. By March 1979 the Psychology Department
had still not cleaned up: cages in the rabbit quarters were
so caked with urine and fecal deposits that "they had to be
individually cleaned with wire brushes, even after six pas-
sages through the cage washer." (Anon.,1979c).

Aside from the discomfort to the animals, such conditions can distort scientific findings. But does anyone
care - except the hundreds of unfortunate creatures being
subjected to these grotesque experiments?

Effects of Brain Mutilation, Starvation, Castration and Isolation

More recently, the same subject has been investigated at
Rutgers University, New Jersey, in mice whose olfactory bulbs
(in the brain) had been surgically destroyed, thus eliminating
their sense of smell. A new wrinkle was added to the study
by depriving some of the animals of food and seeing how
this affected their aggressiveness. (Fortuna,M.,1977).

At the State University of New York, Oswego, 116 male

rats were used to observe the effects of castration on their
killing behavior. Castrated adult rats, some injected with
male hormone, and others not castrated, were each placed in
cages with three one-day old rat pups. The killing response
was defined as "killing and cannibalizing at least one of the
three pups." The different numbers of pups killed by the
various groups of rats were noted. (Rosenberg,K.,1974).

At State University of New York, Buffalo, the effect on
mouse killing in rats was studied after operations destroying
the sense of smell either by cutting the olfactory nerves or
by removing the olfactory centers in the brain. Rats with the
severed nerves, and the unoperated ones, left the mice in
their cages alone; those deprived of olfactory bulbs killed
the mice by "emotional, disorganized biting all over the
body," in contrast to the efficient, quickly lethal biting of
natural killing. (Spector,S.,1972).

A similar experiment, using castrated mice isolated for
30 days to "induce aggression" and encouraged to attack
spayed female mice smeared with urine, was performed under the
auspices of San José State and Chico State Universities in
California by D. Duvall and associates. The urine had been
collected from mice who had been treated with the sex hormones
testosterone and progesterone; and by such methods as count-
ing the number of bites on the victims per fighting bout, the
investigators tried to assess the attack-eliciting characteristics of the various urines used. In the discussion the investigators noted results which were contrary to those obtained by Lee and Griffo as reported above and in another paper
(1976). The crux of the matter seemed to be whether the
attack-eliciting urine smeared on some of the mice was stale
or fresh. Unfortunately, the urine Duvall *et al.* used was

both stale *and* fresh, since all the urine specimens collected
over a period of 20 hours were inadvertently mixed together,
whereby the nuances of stale versus fresh "urinary stimulus
configuration," by the experimenters' own admission, "may
have been lost." (Duvall,D.,1978).

Reader! a penny for your thoughts....

Monkeys Tortured with Shocks Humans
Found Unbearable

In the 50's and 60's, much work on "pain centers" and
"pleasure centers" in the brain was carried out at Yale by
José Delgado, a picaresque neuropsychologist who would like
to see "human society 'psychocivilized' through brain stimulation and other influencing techniques." (Lausch,E.,1972,p.
121). Some of his experiments involved electroshocking monkeys in brain areas known to produce intense pain. (Delgado,
J.,1966).

An idea of the pain suffered by these monkeys may be obtained by comparison with experiments in which the same brain
areas were stimulated in man by B. Nashold and other neurosurgeons. The surgeons implanted electrodes in patients in
the course of operations and recorded the sensations reported
by the latter during electrical stimulation. In the area of
the central gray matter at the level of the superior colliculus in the midbrain, stimulation produced bodily sensations of
pain, burning, vibration and cold. In addition, a patient
would experience feelings which were "described as 'fearful,'
'frightful,' or 'terrible,' and he would become apprehensive
and not allow further stimulations." (Nashold,B.,1969).

Delgado implanted an electrode in the same place (the
central gray area near the superior colliculus) in the brain
of a rhesus monkey named "Harry." First, with the monkey in

restraint, 1.2 milliampere shocks were given. Harry tried to grab anything in reach, bit the chair after every shock, then, released into a small cage, bit the swing as the shocks continued, attacked other monkeys, and finally climbed the wall and clung there.

Although humans, once they had experienced it, could not stand a second stimulation in this brain area, Harry was subjected to 40 minutes of intermittent shocks and 2 minutes of continuous shocks. Other monkeys shocked in the same area responded with screams, grimaces and "signs of aggressiveness directed against the investigator." Through multiple electrodes implanted in their brains, Delgado tested various areas known to cause offensive-defensive reaction when stimulated, and over a period of two days he subjected the monkeys to more than 120 stimulations at each cerebral point, after which they were killed and autopsied.

Although he proposes a hypothesis that brain mechanisms for perception of pain and aggressive behavior have different anatomical and physiological systems, he admits that the evoked aggression was often secondary to electrical excitation of the pain pathways, and that the monkeys demonstrated this by their screams ("high-pitched vocalizations"), dilated pupils, snarling expressions and efforts to escape.

Cat-Rat Fights and Human Violence

The research on prey-killing has been carried forward at Yale University in recent years by John P. Flynn, Professor of Psychology, and his associates Y.C. Huang, C. Chi, R.J. Bandler, and M.H. Sheard. One group of experimenters, for example, took ten cats "which did not spontaneously attack or kill rats," drilled holes through their skulls and inserted electrodes into the ventral midbrain. "With the onset of stimulation the cat went directly to the [anesthetized] rat and bit it repeatedly about the head and neck often with fatal results." (Chi,C.,1976).

The work at Yale has received handsome grants from the U.S. Dept. of Health, ·Education ·and Welfare and several of the National Institutes of Health, and a laudatory mention of Dr. Flynn's experiments in the Dec. 1976 issue of the *Yale Alumni Magazine.* The writer informs the alumni that these experiments "may help man to master his own violent instincts" even though "the 'emotional' aspects of aggression in animals remain beyond the reach of empirical science." (Tucker,J., 1976,p.36).

What also remains beyond the reach of *this* kind of science is the emotional aspects of aggression in the experimenters themselves.

The story of Roger Ulrich's experiments, and their aftermath, discussed below, unhappily confirms this observation.

The Ulrich Experiments

As part of her testimony on Sept. 30, 1976 before a subcommittee of the U.S. House Committee on Agriculture, Christine Stevens read several paragraphs which had appeared in my *Painful Experiments on Animals* (1976), on the work of Dr. Roger Ulrich. A year and a half later, Dr. Ulrich wrote a letter to the Editor of *Monitor*, a publication of the American Psychological Association, which read in part:

"I noticed in the article 'Animal Research: Open Season on Scientists' (Aug. 1977 *Monitor*) that in testimony before a House subcommittee, Christine Stevens, secretary of the Society for Animal Protective Legislation, singled out some of my earlier research on pain and aggression as an example of inhumane treatment of animal subjects. I am heartened that this has happened....
Initially my research was prompted by the desire to

understand and help solve the problem of human aggression but
I later discovered that the results of my work did not seem
to justify its continuance. Instead I began to wonder if
perhaps financial rewards, professional prestige, the opportunity to travel, etc. were the maintaining factors and if we
of the scientific community (supported by our bureaucratic
and legislative system) were actually a part of the problem....

When I finished my dissertation on pain-produced aggression, my Mennonite mother asked me what it was about. When
I told her she replied, 'Well, we knew that. Dad always
warned us to stay away from animals in pain because they are
more likely to attack.' Today I look back with love and respect on all my animal friends from rats to monkeys who submitted to years of torture so that like my mother I can say,
'Well, we know that'...."

It seems only fair to Dr. Ulrich to print this statement
about his change of heart, especially since I was not aware
when I previously described his research that self-doubts
about his role as a behaviorist and conductor of aggression
research on animals had started in the late 1960's, and that
by 1973 he was writing critically: "We studied a single rat
or a single pigeon in a small chamber over a long period of
time...We weren't necessarily interested in the organism; we
were interested only in specific responses of that organism."
(Ulrich,R.,1973).

This inability of animal behaviorists to see animals as
individuals, indeed, as members of a whole society of beings,
has frustrated their efforts to apply their findings to the
human condition. Ulrich is keenly aware of this and is trying to put that awareness into practice. But to appreciate
his present viewpoint it is necessary to know what he has done
in the past (and what many others who have not had a humane
or philosophical awakening are still doing). For this reason,
the following paragraphs from my description of his 1966 article, "Pain as a cause of aggression," are reproduced. (Ulrich,
R.,1966).

"Ulrich's work since 1962, and up to recent years at Western University in Kalamazoo, consisted largely in causing
pain to rats and observing the resulting aggressive behavior.
The investigator would give painful foot shocks to the rats
through an electrified grid floor, with a frequency of up to
38 shocks per minute, or sometimes even higher so that the
shocks were virtually continuous. The intensity of the shocks
(and each intensity lasted for periods of at least 10 minutes)
ranged up to the very strong and painful 5 milliamperes.
'Prolonged exposure to shocks of 5mA. often resulted in paralysis of one or both of these subjects.' A more sensitive
strain of rat (Wistar) could not stand even half this intensity, and four died after exposure to 2 mA. As for duration
of the shock sessions, 200 shocks of various duration 'were
given to six pairs of rats each day for 12 days.' Another
pair of rats were given no less than 15,000 shocks in a period
of 7.5 hours. Another five rats were shocked every day for 80
days, causing them to fight 'more viciously, often cutting and
bruising each other severely.
     Ulrich then introduced other distressing stimuli. The
metal floor of the cage was heated, causing the rats to jump
about, licking their feet as it grew hotter. Then the floor
was cooled with dry ice - this was not effective in producing
fighting: the rats lay on their backs to escape the cold.
Bursts of intense noise (135 dcb., sustained for more than 1
min.) were introduced. The effects of castration were tried;
the animals were shocked wearing hoods, and, finally, one pair
had their whiskers cut off and were blinded by removal of
their eyes." (Pratt,D.,1976,p.61-62).

Scientists Face Their Own Aggression

     The horror, and tragedy, of these appalling experiments
is heightened by the knowledge that some of those who partici-
pated in them now feel that they were scientifically useless.
The *Monitor* article mentioned above stimulated a member of the
Psychology Department at Western Michigan University, Dr.
Robert Brown, to write to Mrs. Stevens in the fall of 1977.
His letter, in part, follows:

     "Dear Christine Stevens,...I am a graduate student at
Western Mich. Univ. studying under Dr. Ulrich, and I thought
you might be interested to know about some of the changes he

has gone through, particularly in relation to the use of animal subjects in research.

Dr. Ulrich, myself and others working out of the Behavior Research and Development lab no longer believe that the scientific information derived from the type of experimentation we previously conducted merits the imprisonment, torture and extermination of any member of any human or non-human species, and that continued research of this sort should stop. We have come to these conclusions based partly on the results of the animal research itself and extrapolations to human behavior. This has led us to a realization of the vital bond between humans and animals, and the necessity of treating animals with the same consideration we show towards members of our own species. Consequently, we are seeking alternatives to the inhumane and wasteful practices of some current experimental procedures....Sincerely, Robert Brown." (Brown,R.,1977).

It would lead us too far from the theme of this book to explore the conversion of Roger Ulrich in detail. However, this animal behaviorist whose experiments admittedly subjected animals to torture now speaks of "my animal friends from rats to monkeys," and has been influenced by the pantheistic philosophy of the American Indian mystic Rolling Thunder, a man who felt that the Earth was an organism and that he himself, and "the deer, snakes, bees, mosquitoes, ants and pinyon trees were one being." (Boyd,D.,1974,p.166). At the same time Ulrich warns against an image of a "new Roger Ulrich" repudiating his past. *The past,* he says, *is to be learned from.* "I don't see myself as doing research anymore but, rather, consider Roger Ulrich a subject and part of the experiment. Under those circumstances, I hardly know who to thank other than the Great Spirit of life that is responsible for all." (Ulrich,R.,1979b).

One of the central events in Ulrich's progress toward this new orientation was the discovery that in the experimental community where he and a group of fellow "behavioral engineers" went to live in the 1960's, they were unable to control

themselves the way they did their animals in the laboratory,
or to solve problems of aggression, distrust, and choice of
goals. What they had learned in the behaviorist's laboratory
was of no avail. Ulrich comments on the behaviorists' approach as follows:

"When we study the individual's behavior, we are studying
a system of relationships; yet, in a sense, we are examining
it too closely. All we see are atomic events, and we overlook
the integrated systems which would explain the behavior sensibly if we could only see it. Our scientific methods of
description suffer from a defective conception of the individual. The individual is not by any means contained inside a
given envelope of skin. This individual organism is the particular and unique focal point of a network of relations which
is ultimately a 'whole series.'" (*Ibid.*,p.42).

By attempting to study aggression in animals "atomically,"
by concentrating merely on the "specific response" of a rat or
a pigeon, by his inability to see the animal as part of a network of relations with other pigeons and rats, and as a part of
a "whole series" which includes himself provoking the aggression, the experimenter fails to get the object of his study
truly into focus. Not only is he unable to understand the
nature of animals; he perpetually fails to learn anything
about "aggression" as it applies to his own kind.

The Face in the Mirror

At the Bronx Zoo in New York the visitor passes through
various animal houses then, turning a corner, confronts a
frame over which is inscribed: THE MOST DANGEROUS ANIMAL ON
EARTH. In the frame, which holds a large mirror, he sees
himself. The experimenter who approaches his animal victims
with scalpel and electrode is also looking into such a mirror.
He can learn little from it about the animal. But if he looks
intently into the glass, as Ulrich finally did, he will dis-
cern there his own menacing image. Surely he can learn much

more by thinking about the torture, both mental and physical, which from time immemorial he has perpetrated on his own kind, than he can learn from artifically-stimulated aggression research on animals. All of us, in fact, must share this burden of introspection, because we are all involved, at different levels of consciousness, in aggression toward our own kind and every other species.

3. DEPRIVATION AND ITS EFFECTS

Deprivation experiments - critics have called them "deprivology" - are designed to interfere with the special senses or the brain centers controlling them, or otherwise to frustrate instinctive behavior ("survival behavior" from the evolutionary point of view). The interference may disturb seeing, grooming, eating, smelling, mating or seeking maternal or peer contact (i.e., protection). The resulting functional disabilities and somatic pathology can be minutely described, but whether the knowledge gained at such a great price is likely to lead to the saving or prolonging of life, or the alleviation of suffering, in humans or animals, is a matter which readers must decide for themselves.[1]

Blinding

One of the first experiments which came to my attention after I became actively interested in animal protection was performed by two zoologists at the University of Texas, Austin, and reported in 1971. (Underwood,H.,1971). The sex glands of male sparrows change in weight in response to

---

[1] *For deprivation experiments specifically on primates, cf.p.128-133.*

seasonal changes in sunlight. The investigators wished to
find out if the stimulus for this was mediated by the retina,
so they took 414 wild house sparrows and blinded half of them
to see if this affected the gonadal response. It didn't. We
are not told what happened to these little victims of science,
but I have wondered what kind of zoologists could live com-
fortably among birds they have blinded. Especially if "His
eye is on the sparrow...."

It appears that the blinding of cats and monkeys is currently a popular experiment. This is usually done by sutur-
ing the lids together, either of one eye or of both, during
the first few weeks of life. The animal is allowed to live in
this condition for months or even a year or two before being
killed and coming to the autopsy table. At that point, brain
sections are prepared for microscopic study, and degeneration
of cells in the visual pathways are carefully mapped. If an
electron microscope is employed, very minute changes can be
detected, so there is always something to report.

Table I, on p.49, from a paper entitled "Effect of lid
suture on retinal ganglion cells in *Macaca Mulatta* [rhesus mon-
key]" (Von Noorden,G.,1977), summarizes the experimental data.
All the monkeys had one eye sutured except No. L35M, who had
both eyes sutured for 10 weeks. The eyes of CR2, the normal
control, were not touched.

Animal No.	Age at lid closure (weeks)	Duration of closure	Age of sacrifice (years)	Remarks
CR 2			Adult (age unknown)	Normal control
L35M	2	10 weeks	2	Both eyes sutured
L150	5	2 weeks	2.5	
F284K	2	12 months	3.75	
L10	4	2 weeks	2	
D21J	2	12 months	6	
7254	9	12 months	6	
B43	2	24.5 months	8	
E291J	1	20 months	2.5	Variable retina shrinkage, data not included

Table 1. Summary of lid suture data
(Von Noorden,G.,1977).

Cats are frequently used. As Loop and Sherman say in their paper,

"A great deal of current research has been directed at an analysis of early deprivation upon the developing mammalian system. A common experimental subject has been the visually deprived cat in which deprivation is produced by eyelid suture. This procedure deprives developing visual system structures of normal light and pattern stimuli and causes a variety of anatomical, physiological and behavioral abnormalities."

These experimenters sewed up one eye in a single kitten, and both eyes in five kittens, before the animals were 10 days old, and "lid closure was rigorously maintained throughout the first 6 months of age." Various tests were made to see how the cats discriminated between light and darkness while their eyes were still sutured and then after the lids were opened. Their ability to discriminate was affected by six variables: open lids; closed lids; dilated pupils and increased retinal sensitivity in the dark; constricted pupils and decreased retinal sensitivity in the light. (Loop,M.,1977).

In a Harvard Medical School experiment, 9 kittens had one eye sutured and 17 both eyes sutured during the second week of life, and the eyes were left sutured until the animals were killed, some after one or two years of sightless existence. Electron microsopic studies of segments of the brain cortex innervated from the closed eye(s) revealed the absence of intracellular structures known as "polysomes" normally found in cells there (in the spiny stellate neurons). (LeVay,S.,1977). Thus, if function is inhibited, structural changes follow, just as immobilization of a limb is followed in time by wasting of the muscles.

How do animals behave after they have been subjected to these mutilations? Sakakura and Doty at the Center for Brain Research of the University of Rochester, New York, working with monkeys, noted the following:

"The blinding seemed to have the effect of 'calming'
these highly excitable wild animals....However, the paucity
of movement and lack of appetite suggest a physiological effect....It was common to discover the blind squirrel monkey duri

the day, unlike the intact animal, curled up in its home cage
in the full posture of sleep, i.e., with its head between its
knees, tail over its shoulders."

In this experiment, electrodes had been implanted in the brain
cortex to be used later for stimulation and electroencephalo-
graphic recording, after which the animals - macaque and
squirrel monkeys - were blinded, two by having their eyeballs
removed, and the rest through having their sight destroyed by
induced glaucoma or photocoagulation. (Sakakura,H.,1976).

Alternatives?...to What?

If an experiment seems to be entirely without significance, it is pointless to seek for alternatives. There are
many experiments on animals which have produced information
of value to humans or other animals; usually this means that
the question which the experimenter hopes to find an answer
to through his research is important and well formulated. But
I do not perceive many significant questions posed by these
experimenters who are engaged in sewing up eyelids and enu-
cleating eyeballs. Loop and Sherman ask: "Can a lid-sutured
cat respond to any light stimulus while still deprived (i.e.,
can the cat 'see' before and/or immediately after the eyes
are opened)?" These experimenters, with their monocularly and
binocularly-sutured cats, have created a highly artificial
situation: if their question had had any value to the human
blind or had needed to be asked in actuality, the answer
would long ago have been found through examination of humans.
Also, since blindness is so widely distributed among humans,
and has always aroused sympathy and exceptional financial support, there is certainly a vast amount of research utilizing
the human blind, both as whole subjects and as operative or
postmortem donors of ophthalmic material. For these reasons,

there seems to be minimal justification for the distressful,
mutilating experiments on animals reviewed above.

Interference with Grooming

An investigator at the University of Oregon raised six
litters of mice, and after one day of life cut one forelimb
off 12 of the mice and both forelimbs off another 12.   In
spite of the amputations, the mice managed to groom their
faces by coordination of shoulder, tongue and eye, only occasionally assuming "exaggerated 'tucking' posture as if they
were attempting to reach the face with the limb stubs." This
was recorded on film over a period of 5 months, and was said
to demonstrate that grooming is not dependent on "exogenous"
contact between paws and face, but is under internal, genetic
control. The National Institute of Mental Health partly fi-
nanced the research. (Fentress,J.,1973).

Several 1977 experiments on grooming in cats after brain
centers controlling this behavior had been destroyed are summarized by J. Diner in his *Physical and Mental Suffering of
Experimental Animals*. (Diner,J.,1979,p.59). The Universities
of Princeton and Iowa participated, and the open-handed National Institute of Mental Health came forward with financial
support. Lengthy film recordings and elaborate statistical
analyses were made, and it was determined that these brain-
damaged cats had been transformed from their usual clean selves
into "filthy" creatures. They had trouble removing tapes
stuck to their fur, licked in mid-air instead of over the surface of their bodies, and failed "to exhibit the normal temporal pattern of grooming behaviors." (Swenson,R.,1977).

These many hours of film showing cats vainly striving to
lick and unstick themselves must be of absorbing interest.
Bravo National Institute of Mental Health for helping to

preserve for posterity this monument to stick-to-itiveness.
I don't use this phrase facetiously: psychologists have been
sticking tenaciously to grooming experiments ever since 1892,
when the German experimenter F. Goltz described abnormal grooming in a dog after bilateral removal of the cerebral hemi-
spheres. (Goltz,F.,1892).

Food and Water Deprivation

W.H. Moorcroft and his colleagues deprived 81 rats of
food and water and measured their spontaneous running about
as compared with control rats which were not starved. The
starving rats' locomotion reached - in 15-20 days - a level
of frenzied activity 10 times that of the normal rats, before
it began to decline, ending with the death by thirst and hun-
ger of the 81 subjects (some lasted as long as 30 days).

The hyperactivity was related to an arousal of that part
of the brain known as the "reticular activating system."
(Moorcroft,W.,1971).

Alternatively...

In more recent work, Kim and Pleasure at the University
of Pennsylvania School of Medicine point out that the effect
of malnutrition on the development of the central nervous system in the whole animal is obscured by many responses (hormonal
among others) to the dietary manipulation. As an alternative,
they have grown newborn mouse cerebellar tissue in culture,
and studied the effect of serum deprivation on the myelin
sheaths of the nerves and the junction points (synapses) of
the nerve endings. The serum-starved nerves were seen to be
retarded in their development, thus reproducing in the test
tube the effects of malnutrition. (Kim,S.,1978).

*Deprived of the Sense of Smell and Sexually De-
sensitized: a Study in "Unnatural" History*

The experiments at the American Museum of Natural History
in New York by L. Aronson and M. Cooper on cats, beginning in
the early 1960's and continuing until Dr. Aronson's retirement
in Aug. 1977 included the surgical destruction (by electroly-
sis) of the olfactory areas, or "bulbs," in the brain and thus
the cats' sense of smell. They also included studies of the
desensitization of the glans penis, genital desensitization of
the female cat, castration of the male, surgical destruction
of the "genital representation areas" in the cortex of the
brain - and the evaluation of all these mutilations on sexual
performance of the male cat.

The ablation of the olfactory bulbs in the rhinencephalon
(the "smell brain") has been mostly shown by other investigators to cause a decrease in mating ability in rodents, but in
the cat, to Aronson and Cooper's surprise, small but significant increases in overall sex behavior occurred. However, the
authors admit that all the efforts that have been made to investigate the effect on sexual behavior of depriving an animal of the sense of smell have been confusing. For example,
experimenters have attempted to produce a loss of the sense
of smell in rats, hamsters and mice by 1. damaging the nasal
epithelium with zinc sulphate, 2. damaging it mechanically,
3. anesthetizing it with procaine hydrochloride, but "the results of these experiments are contradictory. Larsson, Lisk
et al., and Doty and Anisko found that dysfunction of the
olfactory epithelium produced decrements in sexual behavior
similar to olfactory bulbectomy, whereas Powers and Winans and
Rowe and Smith found no decrements at all....In male rabbits,
Stone and Brooks reported continued mating following

destruction of the olfactory bulbs." (Aronson,L.,1974b).

Although they emphasized that "past studies on this problem have all had serious methodological deficiencies resulting,
at present, in a very confused picture," Aronson and Cooper
nevertheless applied to the National Institute of Child Health
and Human Development for a further five year grant, 1974-79,
totaling $195,560, proposing to destroy eight other areas in a
section of the brain (the limbic region, related to the olfactory bulbs) by producing electrolytic lesions there. Despite
failures of "all" past studies, they said that they anticipated
"no difficulties in clarifying the hypersexuality problem and
the relationship of the limbic system to sexual behavior generally." They did admit, however, "the limitation of the lesion
methods in neurological research: small differences in the location and extent of lesions will increase variability, and it
is difficult to separate this variability from other sources of
variability....Lesions of some areas especially the septum may
result in highly aggressive animals that are difficult to handle. This, of itself, may cause decrements in sexual behavior
even if special cages designed to test these animals with minimal handling are used." (They asked for $700 for the construc-
tion of such cages). (Aronson,L.,1974a).

Public Protests Stop the Experiments

This program was suddenly interrupted in 1976 by a wave
of anti-vivisectionist objections, stimulated by a New York
teacher, Henry Spira, who obtained information about the research through the Freedom of Information Act. The protests
spread nationally, fueled by the outrage of friends of cats,
public demonstrations outside the Museum, and wide coverage in
the media. There was a Congressional hearing, loss of face and
of some revenue by the American Museum of Natural History, and,

finally, an announcement in 1977 by the beleaguered Museum that
Dr. Aronson would retire.

However, before that event, he and Dr. Cooper managed to
complete one part of their exploration of the limbic area of
the brain, placing electrolytic lesions in the amygdala. This
seemed to change a male cat's sexual preference, or perception:
in the presence of a female cat and rabbit, he mounted the rabbit. When Congressman Edward I. Koch was told about this, he
was less impressed by its possible scientific significance than
by his discovery that the Aronson-Cooper experiments had cost
the government a total of $435,000.

In Aug. 1977 Dr. Aronson finally left the Museum. The cat
experiments were terminated, and H.P. Zeigler and other members
of the Museum's Department of Animal Behavior busied themselves
with such matters as "Brain stimulation and reinforcement in
pigeons" and "Trigeminal deafferentation and feeding in rats,"
(partially supported by the National Institutes of Health).
Presumably they hope that the friends of pigeons and rats will
not begin questioning the humanity of such experiments on the
scale that occurred in the case of the cats.

Cats to the Rescue of "Hypersexuals" and "Hyposexuals"?

The story of the cats reveals several pitfalls in such experiments.

1. *The impossibility of controlling the many variables,
even the physical ones*. As the experimenters themselves ad-
mitted, the placing of lesions produces uncertain results. In
fact, owing to the immense complexity of the brain, surgical
destructive lesions are really gross mutilations; although they
invariably produce some functional loss (a *descriptive* paper can
always be written), no two have identical effects. As more

experiments are done, not clarification but increased confusion often develops.

2. *The interspecies variables.* For reasons of convenience, the rodent has been the favorite subject of many of these experiments. The primate's sexuality has also been investigated (e.g. at Yerkes Primate Center). Now the cat: "The control of hypersexuality and hyposexuality," said Aronson and Cooper, "are urgent clinical problems, and the results of our proposed experiments should contribute in major ways to the eventual solution of these problems." (*Ibid.*,p.35). If these vague entities are "urgent clinical problems" (which I doubt), it seems that they should be investigated directly in humans. Yet these experimenters waved aside the vast gulf that exists between the human and other species in sexual behavior, especially when their particular animal model was concerned, or when they were applying for a grant and were prepared to drag in half the animal kingdom to prove that the research was "representative."

3. *The psychological variables.* Cats driven to a frenzy by septal lesions have "decrements in sexual behavior" which even behaviorists recognize as emotional outbursts. But the many other emotional variables which can affect sexual performance - fear, depression, confinement, presence of other cats or humans, etc. - are often ignored.

Deprived of the Sense of Smell - and of Sight: the Effect on Territorial Aggression

Hamsters, which live in dark burrows, rely on their sense of smell to detect an intruder; the alien odor precipitates an attack. M. Murphy and his associates showed that if the olfactory stimuli are blockaded peripherally or centrally, the aggression is almost completely eliminated (and so is mating behavior). (Devor,M.,1973; Murphy,M.,1970).

Murphy, who is a psychologist at the Massachusetts Institute of Technology, continued the above experiments by blinding
13 hamsters; there were 13 normal, sighted hamsters as controls.
Each of these 26 animals was put in a cage with a non-aggres-
sive, olfactory-bulbectomized hamster, which they soon attacked.
However, the attack of the blind hamsters continued much longer
than that of the hamsters with intact vision. The author's explanation is that the blind could not see the visual signs of
submission, just as in the dark burrow the attack would continue until the intruder had been forced above ground, where
visual stimuli from the latter's display of the black markings
on his chest - an indication of submission - would inhibit the
attacker. (Murphy,M.,1976).

An Alternative: Let in the Fresh Air

Murphy's observation, above, was ingenious and probably accurate, but it could have been made in the field rather than in
the laboratory, avoiding the mutilating operations. R.M. Lockley's *The Private Life of the Rabbit* (Lockley,R.,1965) comes to
mind, a splendid study in depth of a colony of rabbits, their
fights, their matings, and their surprising physiology. Richard
Adams derived much material for his *Watership Down* from this
book, and in the Introduction to the 1976 edition he says,
"From Ron Lockley I learned that rabbits (as Strawberry protests
to General Woundwort)[1] had dignity and 'animality' - the quality corresponding to 'humanity' in men and women." (Adams,R.,
1976).

Some of the confusion among the animal behaviorists referred to by Dr. Aronson might be dispelled if they stepped

---

[1]*Rabbit characters in* Watership Down.

outside and took a closer look at that "animality." Perhaps
even the American Museum of Natural History will let in some
fresh air. In his 1976-77 Annual Report, the Director, Thomas
Nicholson, indicated that Dr. Aronson's successor would be
"someone whose work would place greater emphasis on natural
populations of animals and on field research, as opposed to
physiologically-oriented laboratory research with domesticated
or laboratory-bred animals." (AMNH,1977).

Social Deprivation

To be deprived of one of the special senses isolates an
animal from some part of the environment. It also frustrates
instinctive behavior, or in terms of evolution theory, behavior
which allows the organism so endowed to survive better. But,
important as seeing, hearing, and sexual expression are, the
need for protection or love, particularly from family or peers,
is equally so. Depriving monkeys of this by isolating them from
their mothers, families or peers has been the specialty of Harry
Harlow and his group at the Primate Laboratory of the University
of Wisconsin since the 1950's. (Harlow,H.,1962). They were
stimulated by the work of R. Spitz, J. Bowlby and others who had
studied psychopathological symptoms in human infants and children separated from one or both parents.

The Wisconsin group kept some young rhesus totally isolated
for many months and up to as long as two years. Some, in a cu-
bicle with solid walls, could see no living thing beside themselves - not even the experimenters, who observed them through
a one-way mirror. The latter then attempted to cure the disorders which their isolation techniques had caused by allowing
the isolates to socialize with "monkey therapists" - normal,
playful monkeys either younger or older - or to be reunited
with their mothers. Not surprisingly, the longer the

isolation the more persistent were the behavior disorders.
At other times, surrogate "mothers" - warm, cloth-covered dum-
mies - were offered to the deprived infants, who avidly clung
to them. But in another experiment, to induce a "schizoid-
like" condition, the dummy would suddenly turn ice-cold,
supposedly simulating the human mother who, according to one
theory, induces the disorder in a child by her unpredictable
and contradictory behavior. The child "splits" between the
desire to cling and the desire to escape; it's in a "double
bind." The monkey, too, is doubly bound - to the accepting-
rejecting dummy mother. Its distress is evident.

"A Modified Form of Sadism"

"Depressive" states were produced by solitary confinement,
sometimes for a month and a half in a "well of despair", a
closed box with stainless steel sides sloping inward so that
the animal could hardly move about at the bottom. Harlow himself called it a "modified form of sadism." Or - and perhaps
even more excruciating - the infant would abruptly be separated
from its mother but could still see her through a partition of
clear plastic. Typically, the infant responded to this form of
separation with violent agitation, protest screeching and persistent staring at the mother, followed by lapses into despair
with whimpering, self-clasping, and immobility. (Suomi,S.,1976).

The Wisconsin group feel that the psychopathology they induce in monkeys through experiments such as the above is similar to that in humans. Its expression will be limited by the
disparity between monkey and human behavior and intellectual
level, but providing the symptoms can be "explained" in both
species by, for example, a disruption of social attachment
bonds, then, they maintain, the syndromes are comparable. I
suspect there is wishful thinking here, especially as they admit

that in these experiments they are "not unaware of human data -
quite frankly, they tend to bias our research efforts." (Suomi,
S.,1974,p.24). Certainly their "monster-mothers" and "wells of
despair" can be expected to produce as much terror, grief and
despair as any well-appointed torture chamber, but that the
reactions are analogous to schizophrenia, psychotic reactions
or sociopathology seems most unlikely. On the other hand, psychotic and psychopathic conditions can be accurately identified
and studied directly in humans, so why continue these misguided
efforts to drive monkeys into insanity when so many humans have
already arrived there?

Much has been made of the fact that the deprived monkeys
consistently preferred the cloth mother to one made of exposed
metal, or wire, even when the latter gave them milk. The experimenters thought that this showed that contact comfort,
rather than feeding, or early sexual attachment, was the prime
factor in an infant's love for its mother. But their simplistic interspecies comparisons have been challenged. It has been
pointed out that contact comfort is *more* important in monkeys
than in man, and anyone who has watched an infant monkey cling-
ing tightly to its mother as the latter swings through the trees
can easily understand why. But Harlow is scornful of such observations. "Look, you will never learn the factors that produce depression or other pathological syndromes in the wild....
Sure, you can get some crude information for evening chatting,
and you would have plenty of evenings, but you will never get
definitive data by observation." (Tavris,C.,1973).

However, the hope that extreme laboratory distortions of
an animal's natural ways will produce information either applicable to man, or of relevance to the species itself, is not
likely to be gratified. As we have seen so often in these

behaviorist studies, the variables inherent in such experiments
confuse the results. Harlow's use of the rhesus as model earns
this comment from R.A. Hinde:

> "The relevance of the monkey data for man might have ap-
> peared in a different light had the early experiments been
> conducted with bonnet macaques rather than rhesus. A short
> separation experience in infancy produces immediate distress
> and detectable long-term effects in socially living rhesus mon-
> keys, but not in bonnets kept under comparable conditions.
> Thus, attempts to generalize from a particular animal model to
> man may well lead to false conclusions." (Hinde,R.,1976,p.195).

False conclusions or not, the juggernaut at Wisconsin Primate Laboratory rolls on through the 1970's. A paper entitled
"Effects of maternal and peer separation on young monkeys" appeared in 1976 (Suomi,S.,1976), and 1977 brought forth "Opponent-process interpretation of multiple peer separations in
rhesus monkeys." (Mineka,S.,1977).

An Alternative: the Ethological-Evolutionary Approach

Roger Ulrich's criticism of the behaviorist's method: that
it fails because it sees only atomic events and not integrated
systems can also be applied to "deprivology." In contrast, the
ethologist's approach - direct observation of the behavior of
the animal in its natural environment - particularly emphasizes
the importance of context. To understand an aspect of behavior,
it must be studied as part of the whole repertoire of the animal's actions and reactions, and these in turn must be fitted
into the context of species behavior viewed in an evolutionary
perspective. This is a paraphrase of comments by Mary Ainsworth, Professor of Psychology at the University of Virginia,
discussing papers by John Bowlby and Dr. Harlow's principal assoc-
iate, Stephen Suomi, at a Kittay Symposium. Although she tempers
her criticism with praise, she concludes with a summary of her

own naturalistic studies of human mother-infant interaction in
the first year of the infant's life. It can be read as an effective alternative to work such as Harlow's and Suomi's. A
quotation will illustrate the quality of her observational method, based, says Dr. Ainsworth, on an ethological-evolutionary
orientation originating with Konrad Lorenz. It is not, she
wryly adds, "popular with grant-awarding panels."

"We identified a maternal behavior that we labeled 'tender,
careful holding'....A mother who handles her tiny baby tenderly
and carefully engenders in him a positive response to physical
contact, and in turn this positive response inspires his mother
to affectionate display, which undoubtedly consolidates his
pleasure in contact. But do babies so handled become spoiled
and overdependent and unhappy when not in contact? No. Even
during the first three months of life, babies so handled protest less frequently when put down than do babies with less
pleasant contact experiences. And by the end of the first year
it is clearly the babies who most enjoy physical contact who
are cheerful about its cessation, and then tend to turn promptly to independent exploratory play." (Ainsworth,M.,1976,p.45).

The National Institutes of Health has poured millions of
public funds into the Wisconsin Primate Laboratory - $2,217,821
in 1978 alone (US NIH,1979) - keeping Harlow's group richly supplied with the now endangered rhesus plus all the torturous
paraphernalia of "deprivology." Yet work such as Ainsworth's is
"not popular with grant-awarding panels." What a commentary on
governmental research priorities!

4. BEHAVIOR UNDER STRESS

Since the days of Pavlov, early in the century, investigators have been conditioning animals by punishment, usually
electric shock, or the fear of punishment. They use the word
"aversive" as euphemism for "painful" or "pain-avoiding;" the
ugly sound of the word fear is often softened by writing it

"fear," as if animals don't really know it as man does, and fear
associated with a painful stimulus may be referred to as "Conditioned Emotional Response," or CER ("in the case of shock-
avoidance responding...a CER is held somehow to add to or increase the level of aversive motivation").

An experiment by S.R. Scobie at the State University of
New York, Binghamton, subjected 48 rats to shock which they
learned to avoid in a shuttle-box. After training, they carried
on avoidance behavior when they heard a signal (the conditioned
stimulus) which had been associated with the shock. If a second
shock was then given along with the signal, they became fearful
and ran faster in the shuttle-box. But when this second shock
was made very intense, to produce overwhelming fear, the rats
froze with terror and were no longer able to flee from the
electrified area delivering the original shock. (Scobie,S.,1972)

This is only one of innumerable ways investigators have devised to develop then interfere with an animal's attempt to
avoid a painful stimulus.

Physical Disintegration:
Gastric Ulcers

Jay Weiss at Rockefeller University has performed similar
experiments on rats trained in avoidance at the sound of a warning signal, only to receive a shock every time they execute the
previously correct response. As a result they developed severe
gastric ulceration. Weiss is unusually ingenious in his treatment of rats. He likes to work with pairs, with one partner
able to avoid shock by a manipulation; the other bound and helpless. "The rats were subjected to one continuous stress session
lasting 21 hours, with the shocks, each preceded by a signal,
scheduled to occur at the rate of one per minute. After the
conclusion of the session all animals were sacrificed [the

investigators' euphemism for "killed"] and their stomachs were examined for gastric lesions." On an average, the bound, helpless rat had three times as much stomach ulceration as his "avoidance" fellow. (Weiss,J.,1971).

Cold-Restraint

Another method of producing gastric ulceration in rats, devised by S.C. Boyd and associates at Vanderbilt University, Nashville, combined so many stressful procedures that a canceling-out rendered the experiment partially ineffective. The experimenters used "cold-restraint" combined with electric shock. Rats were starved for either one or two days, then immobilized in plastic restraining devices and placed in a refrigerator, with a temperature of from 3° - $7^{\circ}C$. The animals were left in the icebox for periods up to six hours, and the degree of ulceration was found to vary with respect to drugs administered (reserpine, atropine) and the length of time in cold-restraint. However, when different schedules of painful electric shocks were given, the experimenters could not produce variations in the degree of ulceration.

They decided that this didn't mean that the half-frozen animals had become insensitive to pain, because they squealed and struggled when shocked. But for these psychologists, anyway, the animals' failure to respond with more or less ulceration to classical variations in administering shocks frustrated the experimenters' efforts to investigate "psychological factors."

Reader, imagine yourself terrified, half-starved, half-frozen and immobilized in restraint. You are suffering intensely from gastric ulcers. A psychologist subjects you to painful electric shocks. Can you imagine what psychological material he will be able to elicit from you?

Screams? Struggling? Too trivial to discuss.

Boyd et al. conclude:

"Cold-restraint is viewed as an inappropriate technique to
use in further examination of the influence of specific psychological factors on stress-induced ulceration." (Boyd,S.,1977).

Psychological Disintegration: Learned
Helplessness

C.L.J. Stokman and M. Glusman at the Psychiatric Institute,
New York City, planted electrodes in the brainstem (hypothalamus)
of four cats. "Stimulated" electrically, the cats showed a
flight response. After ten daily sessions of this, the animals
were again stimulated to flight but immediately punished by a
strong foot shock each time they attempted to flee. Eventually,
the punishment suppressed the flight reaction, but some of the
subjects needed much more punishment than others before they gave
in. (Stokman,C.,1969).

This technique of subjecting animals to "inescapable" electric shock was originated by M. Seligman and J.B. Overmier and
first reported in 1967. Dorworth and Overmier of the University
of Minnesota in a recent paper (1977) describe the appearance of
dogs subjected to this treatment. Restrained in a hammock, they
are first given a series of inescapable shocks. Later, they are
shocked in a situation where escape *is* possible, but instead of
running off they fail to become "active and vocal, and react to
shock with whimpering and passivity; they seem to give up all efforts to find a response for coping with the shocks [and] look
pathetically helpless." The phenomenon has been named "learned
helplessness." (Dorworth,T.,1977).

However, if a dog is given only one session of inescapable
shock, the resulting helplessness will dissipate in time. This
is true only for the dog not reared in the laboratory: for laboratory-reared dogs, and for rats, the helplessness does not
dissipate. Seligman and his associates at the University of

Pennsylvania suggest that the dogs of unknown past history have,
in their prelaboratory days, succeeded in escaping from "natural
aversive [i.e., distressing] events," in contrast to the cage-
reared animals who are relatively deprived of escape experience.

"Escaping natural aversive events" could also have been expressed as "having prior experience of kindness and help from
owners." The "36 mongrel dogs" used by Dorworth and Overmier
(1977) were very likely all ex-pets, either abandoned, strayed
or stolen. Apparently they still hoped that the torture being
inflicted on them would be relieved by the experimenters, whom
they doubtless associated in their minds with those other of the
human species they had once loved. The laboratory-reared dogs
and rats knew better.

Seligman and associates discovered one procedure (1968) that
"reliably 'cured' dogs that repeatedly failed to escape in the
shuttle-box following exposure to inescapable shock. These dogs
were dragged by leashes back and forth across the shuttle-box to
'show' them that responding terminated shock. This form of 'putting through' was successful in all cases, but required 20-200
draggings before the dogs responded reliably on their own."
They used a similar procedure on the rat. (Seligman,M.,1975).

The Cause of "Learned Helplessness": Chemical or Cognitive?

Attempts have been made to relate the behavior changes observed under stress to changes in brain neurochemistry. Chemical substances known as "neurotransmitters" can either increase
or decrease the perception of pain or stress, are apparently related to emotional depression and are also influenced by, or
themselves influence, schizophrenic states. It has been suggested that the phenomenon of "learned helplessness" can be
explained by the depletion of one of the neurotransmitters,

norepinephrine. (Weiss,J.,1976,p.157). S. Maier and Seligman, on the contrary, dispute this and relate the phenomenon to the fact that the animals exposed to inescapable shock have, literally, learned that there is nothing to be done; their behavior reflects that knowledge.

Does Teaching Animals to Despair Increase Understanding of Human Psychopathology?

What can be made of all this? Surely these experimenters are contributing little or nothing to an understanding of the complexities of human anxiety or depressive states. If anything, these tortured and terrified dogs appear to be suffering from a traumatic reaction, similar to the soldier's "shellshock," battle exhaustion or gross stress reaction occurring under circumstances in which men felt that physical destruction was inescapable.

Psychiatric study of human reaction to maximal stress has concentrated on the subjective symptoms: anxiety, repetitive nightmares, depression. Animal behaviorists, lacking access to such phenomena as the dreams of their subjects, have concentrated on objective signs, such as "learned helplessness." This concept seems to have captured the imagination of many behaviorists as a very innovative idea. However, helplessness associated with traumatic situations is hardly a new concept to students of human psychopathology. Freud in 1926 observed the significant role it played, noting that helplessness manifested itself physically when the danger was objective and psychologically when it was instinctual. (Freud,S.,1926). In 1948 Franz Alexander wrote that traumatic symptoms "can be explained as signs of damage to the ego, which, under the intimidating influence of the trama, abandons its mastery of coordination and regressively retreats to helplessness." (Alexander,F.,1948).

Unfortunately, there has always been abundant human material of this sort available in our war-torn world, and no lack of
publications on the subject. Kardiner's *Traumatic Neuroses of
War*, for example, published as long ago as 1941 and an influen-
tial contribution to military psychiatry on the eve of the
United States's entering World War II, was a penetrating study
of acute anxiety in humans under stress. (Kardiner,A.,1941).

Yet B.M. Mitruka, writing in 1976 about the "Use of Animals in Behavioral Research," in a textbook of which he is
editor-in-chief, seeks to illustrate some of the psychodynamics
in human manic-depressive psychosis by citing a 1957 experiment
by C.P. Richter, from the Psychobiological Laboratory of Johns
Hopkins Medical School, in which many rats rapidly gave up swimming and drowned when placed in small cylinders of water from
which there was no escape. (Mitruka,B.,1976,p.242). Mitruka
says that these rats are exhibiting "abnormal behavior." Normal behavior would be to continue swimming for several days im-
pelled by an instinct or unconditioned reflex of self-preservation. However, the rats are overwhelmed by their loss of control of the environment, since their efforts are totally un-
rewarding, and sink into "maladaptive passive behavior" analogous to that experienced by humans suffering from a manic-depressive psychosis. They resemble Seligman's dogs, who became
helpless in the face of inescapable electric shock.

"Where There's Hope There's Life"

Mitruka's statement shows how behaviorists are anxious to
relate human psychology to animal models and at the same time
how they distrust feelings - emotions - in animals as determinants of behavior. The animals are said to experience "loss
of control," as if their bodies were vehicles which they are no
longer able to steer; they "learn helplessness." The fact that

they are terrified is passed over. Instead, the rats' behavior
is interpreted as a failure of the survival instinct.

But this is not the explanation given by Curt Richter, who
performed the experiment cited by Mitruka. Although Richter's
aim was also strictly behavioristic - to obtain "constant, reproducible endurance records" of domesticated and wild rats
swimming until they drown - he admits that an emotional reaction, hopelessness, was decisive in their behavior. Richter
forced his rats to swim in jars 30 inches deep but only 8 inches in diameter, with a jet of water playing onto the surface
of the water, in the center, to prevent the animals from floating. Prior to being placed in the jars, the rats were held in
the experimenters' hand, in a black holding-bag. Some of the
rats had their whiskers clipped. A few rats, particularly the
domesticated ones, swam as long as 80 hours before they drowned;
many of the wild rats, who reacted much more strongly to any
restraint, either died while being handled or shortly after being placed in the water; most of the rats drowned quickly if
their whiskers had been removed. Richter believed they all died
from hopelessness, compounded by the loss of whiskers which
destroyed "their most important means of contact with the outside world." (Richter,C.,1957).

Richter connects these sudden deaths with "voodoo death"
among humans. In this, an individual who has broken a taboo is
told by a witch doctor that he must die, and after experiencing
great terror, sinks into despair and, in fact, often does die.
Richter cites other instances of sudden death among apparently
healthy people, for instance among those terrified of an operation, or among young soldiers, all of whom apparently die as a
result of a belief in an approaching doom. In spite of Mitruka's
suggested diagnosis, Richter himself does not draw a parallel

between the psychological disintegration of these rats and a
manic-depressive psychosis.

We therefore return to the point which I made on p.68,
namely that the stress which animals are subjected to in this
type of experiment merely produces a kind of traumatic reaction
which has long been known from observations in humans. However,
the analogy or extrapolation is greatly contaminated by variables in the procedure which have little or nothing to do with
situations in which humans are likely to find themselves. Richter cites respiratory and vascular reactions in the rats - from
being forced to swim in a nearly upright posture, or being held
upright in the experimenter's hand, or being suddenly immersed
in water; these may be as important as fear in producing the
slowing of the heart rate and respiration, and heart failure,
which are the physiological causes of death (for a discussion
of how variables negate the value and ruin the application to
the human condition of many of these animal experiments, see
p.14).

Trustful Rats; Treacherous Humans

There is a final turn of the screw in this macabre experiment. The rats' hopelessness subsided when the experimenters
from time to time pretended to free them, or momentarily "res-
cued" them from the water. After this, imagining that the
situation was not quite hopeless, the wild rats would swim just
as long as the domesticated ones had, the latter having been
more optimistic in the first place because of previous experience of "gentling." This reaction, incidentally, is comparable
to that of Seligman's "mongrel dogs," described on p.67, who
were able to overcome the feeling of helplessness produced by a
relatively brief exposure to inescapable shock if they had had
some prior experience of human support. Richter used this

releasing and taking-up technique to free the rats from feelings
of hopelessness and thereby to achieve those "constant, reproducible endurance records," unaffected by individual emotional
variations, which to him were obviously a much more suitable
subject for behavioral analysis.

But of course this only postpones the rats' knowledge of
their doom. Finally, the treachery of the experimenters becomes
apparent to them, and the exhausted animals give up and drown.

In recent years, Richter appears to have transferred his
interest from rat swimming marathons to study of the animal
biological "clock." To see whether this is independent of the
animals' perception of light and darkness, they are blinded just
after birth. This was done to rats (Richter,C.,1971), and more
recently to chipmunks (Richter,C.,1978).

Others have sustained their interest in rat swimming experiments since Richter's 1957 study: for example, rats have
been forced to swim with weights on their tails, or while suffering from arthritis, or in icy water, etc. (LeBlanc,J.,1972).
And every year *Science Citation Index* lists a group of these experiments under the heading, "Rats; swimming."

An Alternative: College Students
Learn to be Helpless

Curt Richter draws an analogy between learned helplessness
in rats and the conviction which under special circumstances
makes humans feel that they are helpless to escape impending
doom. But if humans are also susceptible to learned helplessness, why not study *them* instead of lavishing time and resources
on rats? This is a good question to put to the National Institute of Mental Health, which liberally dispenses public
funds to those who specialize in teaching animals to despair.
(Diner,J.,1979,p.140).

That it is possible to demonstrate learned helplessness in humans is proved by some harmless experiments by David Glass on college students (and a few older subjects). Glass reports these in a book which explores the relationship between a proneness to coronary disease and certain personality types. Two types, A and B, predominate, characterized as follows:

"Type A individuals exert greater efforts than Type B individuals to master stressful events which they perceive as a threat to their sense of control. These active coping attempts eventually extinguish in the face of uncontrollable stimuli, for without reward the relentless striving and time urgency of the Type A individual leads to frustration and psychic exhaustion, which culminate in giving up efforts at control. An almost ironic reversal of behavior is then observed, with Type A individuals showing greater signs of helplessness than their Type B counterparts." (Glass,D.,1977,p.7).

This is illustrated in an experiment in which both the A's and the B's were subjected to noise bursts of a certain intensity. In the "pretreatment phase," they were told that they were able to control the noise by turning a knob in a particular way; however, unknown to half of the subjects was the fact that their knobs were unconnected with stimulus circuits and therefore for them the noise stress was inescapable. Thus some of them "learned helplessness," and this was reflected, in the subsequent "test phase," by the way these "no escape" subjects approached the task of curtailing renewed noise bursts through manipulating a handle. Their performance proved true to the types differentiated above.

Aside from avoiding the suffering customarily inflicted on animals by such experiments, this one had the great advantage of the subjects themselves being able to describe their feelings of helplessness and lack of control.

*Many Variables Confuse the Animal
Behaviorists' Findings*

As far as psychiatrists and clinical psychologists are concerned, they continue to care for their patients largely unaware
of these behaviorist animal experiments of which they are the
supposed beneficiaries, while those behaviorists who are hooked
on the use of punishment via electric shock pursue their study
of patterns of "punished responding." Drugs, especially tranquilizers, sedatives, anti-depressants, stimulants, and narcotics, are tested to see what effects they have on the rate of
responding. Obviously the behavior of an animal under stress
will be affected: sometimes the rate increases, sometimes de-
creases. If these changes were strictly and invariably related
to the particular drug used, something might be learned, but in
fact the results often end up all over the place because of the
tremendous number of variables which operate in even the simp-
lest experiment. Here are some of them: age of animal, sex,
weight, genetic strain, diet (including duration of starving
prior to testing), temperature, humidity, cage construction,
bedding, time of day when drug is administered, number of test
animals per cage, speed of drug administration, time elapsing
between administration and testing, health of animals, extent
to which animals were handled in infancy as well as prior to
test, variations in drug (manufacture, age, storage, bottling,
etc.), insecticide sprays used in laboratories, skill of technicians, errors of recording, and, as discussed above, the
differences in the experimenters' behavior as perceived and reacted to by the animals.

Further variables are introduced by the use of electricity.
V.P. Houser states:

"The electrical resistance of animals varies widely (i.e.,
from several thousand to several million ohms) as they move

about on a grid floor....Thus grid shock is by its very nature
a variable stimulus. By placing a large resistance in series
with the animal subject, the constant-current shock generator
reduces the variability....This design is limited, however, by
the fact that very large series resistances require the use of
high source voltages. These voltages, in turn, can produce
painful current arcs from the grid to the animal's body."
(Houser,V.,1978,p.71).

Most devastating of all are the effects of the fear and
pain themselves: see Hillman's list on p. 51 , ending with the
comment that stress "is almost certainly the main reason for
the wide variation reported among animals upon whom painful experiments have been done." (Hillman,H.,1970,p.51)

Houser also ends his 80 page review of the subject with a
warning: the properties of clinically active drugs cannot be
understood through their supposed effect on anxiety and fear and
on behavior associated with these emotions. He shows that the
experiments that are likely to produce the greatest distress
("conditioned emotional response" - CER - in the face of unavoidable shock) are just those which are least useful in elucidating
the effects of these drugs. (Houser,V.,1978).

Unfortunately, he arrived at this essentially negative information partly as a result of an experiment which must have caused
the most appalling suffering to the dogs who had to undergo it.
Restrained on a table by ropes and a choke collar tied to an iron
pipe above them, they were trained to escape shock by pressing a
response panel. They then underwent surgery to remove the bladder and to externalize both ureters so that urine samples could
be continuously collected without storage in the bladder. After
recovery, they were subjected to unavoidable 8 mA shocks to their
shaved legs during 140 sessions *from which they could not escape*
- and this went on for an unstated number of days. A tranquilizing drug, chlorpromazine, was then given, but under the stress

of the torturous procedure being inflicted on them, the dogs proved relatively insensitive to it. (Houser,V.,1976).

Humane Refinements Are Possible in Animal Experiments

Let us assume for the sake of argument that the human demand for drugs to control states of anxiety and depression, and to alleviate pain, will inevitably result in a continuation of neurophysiological experiments on animals. Some might feel that animals in a sense have an obligation to contribute to the pain studies, since there are obviously many veterinary uses for pain-relieving drugs. It may therefore be realistic to look for refinements in technique in those procedures in which animals are the standard experimental models, without, of course, relaxing the demand for reduction of the numbers used and replacement by non-sentient material whenever possible.

The use of electric shock is open to such abuse that one hesitates to say anything which might be construed in its favor: "he who sups with the devil should use a long spoon." Nevertheless, as a possible first step toward alleviation of the distress of animals subjected to this technique, I offer a statement from the College of Medicine of the University of Florida at Gainesville. It appeared in the 1976 Annual Report of that facility under the Animal Welfare Act to explain why 10 of the 255 primates used in experiments that year received no pain-relieving drug.

"These animals received painful or non-painful electrical stimulation of their legs, and they *always* had the option of turning off the stimulus if they wished to. They were never subjected to long periods of painful stimulation over which they had no control. The animals tolerated these procedures very well. They have remained in good health with testing over periods of several years and have shown no overt signs of stress. They quite willingly enter the experimental testing situation, and they do not show undue signs of fear of the testing apparatus.

The experiments involving painful stimuli were specifically designed to study central nervous system mechanisms of pain, and, of course, it is impossible to do so without delivering stimuli sufficient to elicit pain reactions." (USDA/APHIS, Florida,Univ.of).

Another experimenter who, without abandoning animal experimentation, nevertheless seems to be searching for more humane methods, is Susan Iversen. She is a research psychologist at the University of Cambridge in England and the co-author of *Behavioral Pharmacology*. (Iversen,S.,1975). Why create a change in behavior (e.g. helplessness) through the application of severe physical and psychological pain, she inquires, when a much milder measure - a low dose of the stimulant amphetamine - will produce a simple change in the rat's locomotion which is equally informative? A drug can be injected to block the amphetamine effect, whereupon the rat's locomotion returns to normal. Iversen concludes:

"In regard to the particular research interests of my laboratory, while it is important to understand the etiology of abnormal behavior in animals and man, behavioral measures which involve psychological stress do not provide a necessary or desirable means of identifying drugs of therapeutic use." (Iversen, S.,1976,p.64).

In other experiments of this type, the normal circadian (around the clock) activity of the animal can be observed for any alterations which may be caused by toxicity; also, variations in feeding habits can be noted. There remain a good many other variables which can effect the results, but the absence of fear and experimental pain will eliminate those which are both the most confusing and the most inhumane. (NRC,1977,p.111).

Still another example is the investigation by Pieper of stimulant and depressant drugs in great apes (six chimpanzees and two orangutans). Before being given the different drugs,

the apes were taught to pull a series of levers in a sequence
cued by lights of varying brightness above the levers. If the
sequence was followed correctly, the ape earned a candy or nut
reward ("reinforcer"); if an error was made the lights were ex-
tinguished and no food was delivered during the seven second
time-out. On certain days a placebo (inactive substance) was
substituted, allowing behavior then to be compared with that on
the days when the apes had received active drugs. The experiment demonstrated that mild "punishment" in the form of being
deprived of a reward is an effective stimulus; certainly it is
much more humane than the electric shocks so often used to punish animals who fail in learned responses. It's even likely
that the apes looked forward to the daily "game" and its rewards. Some of the test drugs may have made them uncomfortable,
but when they experienced frequent failures and consequently
received no reward, they simply stopped responding and the session was terminated. (Pieper,W.,1977,p.167).

An Alternative: Learned Self-Helpfulness

As soon as attention is paid to the type of experiment which
modifies behavior by reward instead of punishment, one can perceive much more creative tasks for the behavioral psychologist.
In one direction his role begins to merge with that of the
ethologist, in another with the science teacher seeking to introduce students to non-destructive observation of animals and their
behavior. The behaviorist's technique of "operant conditioning,"
defined as "behavior maintained by its own consequences," even
has a place in medical rehabilitation, judging by the success
achieved with spastic children taught to use an electromyograph,
an instrument which registers the degree of muscular control
through variations in audible tone. Muscular relaxation is re-
warded by a low tone ("Low tone becomes synonymous with the

therapist's smile"). (Cataldo,M.,1978). This is the technique
of biofeedback, which has proved valuable in alleviating conditions such as poor circulation in the limbs, muscular tension,
epilepsy and various stress-related symptoms. It works through
a combination of self-monitoring of physiological functions
(blood pressure, muscle tension, brain wave activity, temperature, etc.) and an enhanced sense of self-control and mastery.
(Pew,T.,1979). For helplessness, learned self-*helpfulness*.

One of the chief reasons why such a system and other "con-
sciousness-raising" techniques are antithetical to the animal
experimentation I have been reviewing in this chapter is their
emphasis on reflective thought in contrast to the quick, knee-
jerk response exacted by classical conditioning. Animals subjected to electric shock, to a painful stimulus, have little
time for reflection! Animal psychologists, perhaps unconsious-
ly still stuck at the Cartesian animal-as-machine level, seem
to deny their animal subjects the time to think, the ability to
reason and the capacity to feel. It takes an experiment on humans to jolt psychologists of this type out of their mental rut.

For example, Turner and Soloman subjected 59 Harvard and
M.I.T. students to foot shock to the right ankle, painful "to
the limit of the subject's endurance." There were various ways
different groups of students could avoid the shock: one, the
quickest and most reflex-like, a movement of the big toe of the
right foot; another, the slowest and most "reflective," sliding
a knob back and forth. The toe movement group generally failed
to learn that this *was* the way of avoiding shock: the movement
was too quick and reflex-like for them to notice it and they
viewed the shock as uncontrollable. Those with the knob readily
learned that this was a successful avoidance response. (Turner,
L.,1962,p.95).

This experiment has helped investigators to understand why rats and dogs escape from a painful electric shock more successfully when they are allowed to become aware of the connection between their behavior and the avoidance of shock - in other words, when there is time for feedback to be received and to reinforce the avoidance maneuver.

It all seems a long way around. Since we are primarily interested in ourselves, not as "king-sized rats," but as humans, why not cut out the rats and concentrate on the Harvard students (if the latter are still willing)?

Four

CANCER

Our Environment is Killing Us

There is now widespread realization that many if not most
cancers, perhaps 80%, are caused by substances in the environment: cigarettes, fossil fuels, pesticides, drugs, preserva-
tives, food colors, dyes, radiation, asbestos and many others.

Some of this has become known through the detection of
cancer among people who are particularly exposed to these substances, including those engaged in their manufacture. As long
ago as the 18th century, it was realized that the scrotal cancers of chimney sweeps were caused by skin irritation from
soot. Investigations in recent decades have demonstrated that
cancer can be induced in animals by an increasing number of
these substances, including all those which have been proved
to be carcinogenic in humans. But it is not known how many
carcinogenic substances in animals will produce cancer in
humans, nor can it be assumed that different species of animals
are uniformly subject to cancer induction.

*Variables Impair Scientific Value
of Animal Tests for Cancer*

The different susceptibility of species is only the beginning of an almost endless list of factors which affect an
individual animal's chance of acquiring the disease, and
which influence its nature and duration. Among these are sex,

81

in-breeding, age, diet, the combined effect of two or more carcinogens (may be additive or inhibitory), husbandry (impurities
in the diet, water or air, bedding, housing conditions - single
or group caging - crowding stress, etc.), immune status, and the
way the carcinogen is metabolized in a particular species. This
variability is further complicated by a variation in response to
each type of carcinogen: thus the rabbit is refractory to liver
tumor induction from azo compounds but susceptible to skin tumors from polycyclic hydrocarbons, whereas the rat is susceptible to these in opposite fashion. (Grice,H.,1975,p.13).

Proceeding to the matter of the substance to be tested and
the selection of dose and frequency of exposure, another Pan-
dora's box of variability opens. Assuming that the objective is
to extrapolate (transfer) to humans the cancer potential demonstrated in animals, there must be a relationship between the
conditions under which both humans and animals are exposed to the
substance being tested. Saccharin, for instance, fed in enormous doses to rats, produced bladder cancer, yet this hardly
mimics the typical human exposure to comparatively minute quantities of the substance, especially when it is likely that such
massive doses may overwhelm the natural defense mechanisms of
the body, which normally can be counted on to defend against or
even repair potentially cancerous changes in the genetic material
of the cells. (Gehring,P.,1973).

Another problem is the frequency and duration of exposure
to the test chemical. In order to be comparable to the very long
development period of most cancer, the material may have to be
administered daily over the life span of the test animal - and
cancers are also very slow to develop in animals: in monkeys,
the species closest to man, up to 10 years is required to develop
tumors, and 5 to 10 years is not uncommon in dogs. Thus the investigator is forced to concentrate on rodents, but even with

these a year or two may be required, during which all the
conditions which may affect the experiment must be kept constant,
and, if there are results worth reporting, these are still in
species far removed from our own.

Finally, the impurity of the test substance can skew the
results. With the increasing sensitivity of procedures such as
the Ames *Salmonella* Test, a mutagenic (or carcinogenic) effect
can be demonstrated in what was supposed to be a harmless substance simply because it was contaminated by a tiny amount of a
potent carcinogen. In fact the saccharin in the experiment described above was demonstrated by the *Salmonella* test to contain
traces of just such a substance, which might be another explanation for the cancers which appeared in the rats. (Donahue,E.,
1978). The Grice study noted that "factors such as batch to
batch variation, chemical synthetic processes, packaging, handling, storage and processing into food and drug preparations and
interacting with other chemicals can have profound influences on
the toxicity of the test material. Long-term studies should not
be undertaken until the above information is available and un-
equivocally established." (Grice,H.,1975,p.20).

Unfortunately, the emphasis on the use of animals in bioassay of slow-developing cancer has resulted in vast numbers of
such studies, all subject to the variable factors described
above, and therefore frequently of questionable scientific value.
Furthermore, if the variables were uncontrolled, or unrecognized
and not reported, it is impossible for a later investigator to
reproduce the experimental results. It is disturbing to think
of the wastage of animals in these investigations, the loss of
human time, and the expense.

Nevertheless, the high cost of carcinogen investigation
in animals - an experiment involving typically 600 to 700 animals and costing more than $100,000 (Sontag,J.,1976) - and the

disappointing results in terms of prevention or cure of the
disease in humans, have raised a clamor for a fresh approach.
Humane considerations have had little to do with it, but humanitarians can nevertheless find encouragement in the new trend.

How Cancer Cells Get Their Start

In the light of recent developments in cancer research and
better understanding of the molecular and cellular mechanisms
involved, it has become retrospectively clear why the animal
testing of the past came up with so few clues to the nature of
the disease in humans. A brief review of the mechanism of
carcinogenesis as now understood will help to explain this.

It is now believed that the chemicals which act as carcinogens must be activated or potentiated in the body by enzymes
which turn them into "electrophiles" - compounds which "love"
electrons and which seek them out within the cell nuclei from
DNA, which carries the genetic information for all cells, and
RNA which transmits information from DNA to the protein-forming
system of the cell. In the process of raiding DNA and RNA for
electrons, the electrophiles become bound to DNA and cause a
change or "mutation" in its genes. The genetic message is
disrupted, and the result may be the initiation of an abnormal
or cancer cell. Next, a second substance, or "promoter," which
may be a hormone, causes the initiated cells to clump or pile up
and grow chaotically into a cancerous tumor. The enzymes which
help to start all this trouble can vary in different individuals,
and even more so between species, thus explaining why humans
differ one from another in susceptibility to cancer and, by the
same token, differ even more in this respect from other species.

Obviously, if confusing variables increase when the eye
of the scientist strays from human material, it behooves him to

look to his own kind for some of the answers. Here are some of
the techniques which are being developed.

Alternatives: Human Cell and
Organ Cultures

Since experimental cancers cannot be safely initiated in
human beings, investigators have turned to the study of human
cells, tissues and even minute parts of organs which can be
kept alive in the laboratory if nourished by a suitable medium.
"*In vitro*" studies of animal tissues have also played their
part; indeed in certain circumstances are still valuable, but
for the reasons mentioned above the hope of the future seems to
lie more with the techniques which use human material to understand humans. One of the breakthroughs occurred with the development of cultures of human embryonic lung cells by Leonard
Hayflick, reported in 1961. (Hayflick,L.,1961). These "WI 38"
cells are able to produce forty generations before dying. However, the cells are fibroblasts, which differentiate into the
connective tissue of the body (tendons and the like), whereas
most cancers are derived from epithelial cells. Much effort
has gone into attempts to culture epithelial cells from various
organs, also to grow and keep alive for longer periods of time
cells which were not merely embryonic but more differentiated.

Important advances along these lines were discussed at
a workshop at the Given Institute of Pathobiology, Aspen,
Colorado, in Aug. 1977 sponsored by the Carcinogenesis Program
of the National Cancer Institute. (Harris,C.,1978). Some of the
reports of the symposium are summarized in the following paragraphs.

The secret of long life for cells *in vitro* has been found
in the addition of specific substances to the culture medium,
replacing as much as possible the traditional serum whose large

protein content interfered with growth and which was subject to
deterioration. Thus the addition of "epidermal growth factor"
has tripled the culture life of epithelial keratinocytes from
foreskins of newborn infants; cultured human mammary epithelial
cells have been maintained with a mixture of hormones (estra-
diol, progesterone and prolactin), and it is now thought that
each epithelial cell type may require a different mixuture of
hormones.

Significant advances in organ cultures or explants were
discussed at the Given Workshop. It was reported:

"Bronchus, pancreatic duct, bladder, uterine cervix,
breast, colon, esophagus and aorta can all be cultured from
weeks to months with maintenance of normal-appearing epithelium.
Chemically defined culture media have been developed for human
bronchus, colon and pancreatic duct, which eliminates the biological variability due to serum from studies of carcinogen
interactions in these tissues."

Cancerous Transformation of Normal
Cells in Culture

A major topic at the Given Workshop was the metabolism of
chemical carcinogens by human cells and tissues. The inter-
individual variation in binding levels of the carcinogenic polynuclear aromatic hydrocarbon BP in liver biopsy specimens varied
30-fold, in cultured human colon and bronchus 60 to 70-fold.
These variations in an individual's "metabolic profile" may be
influenced by genetic factors, disease states and the environment. Eventually, as Dr. U. Saffioti suggested at a New York
Academy of Sciences meeting in June 1978, tests on biopsy material from individuals may help identify those with particular
susceptibility to certain carcinogens and warn them to avoid
them (e.g. cigarettes). Similarly, they might enable an
industry manufacturer using a potentially hazardous chemical to

exclude susceptible individuals, identified by an appropriate
test, from the workplace. (Edelson,E.,1978).

In addition to reports on the metabolism and DNA binding
of known carcinogenic chemicals by cultures of human bronchus,
colon and arteries, successful transformation of normal human
cells from the mouth (fibroblasts) and foreskin into cancer
cells was described. These last-mentioned experiments are
similar to others (not reported at the Given Workshop) in which
A.E. Freeman and others (Freeman,A.,1977) described a chemically
induced cancer-like change in human skin cultures, and R. Tayler
and D.W. Piper (Tayler,R.,1977) reported that cigarette smoke
condensate produced a result on the stomach mucosal cells in organ culture typical of malignancy that did not differ from that
of the known gastric carcinogen N-methyl-N'-nitro-N-nitrosoguan-
idine. With this research we are approaching a very significant
stage in alternatives to cancer testing in animals: namely, the
ability to produce, in the laboratory, cancers in human organs
in response to chemical substances known to be mutagenic or
carcinogenic.

The Ames *Salmonella* and Cell Transformation Tests for Mutagenicity

The knowledge that a substance is "mutagenic" (capable of
producing a heritable change in the genetic material of a cell)
does not have to come from trial and error testing of innumerable
environmental chemicals in delicate cell or organ culture systems. It is very cheaply and quickly demonstrable (for about
$200 and in 3 days for each assay) by the Ames test, using
Salmonella bacteria which cannot grow because a mutation has
made them unable to manufacture the amino acid histidine, a
necessary nutrient. The potential carcinogen to be tested is
exposed to rat or human liver extract and, just as would happen

in the mammalian body, is converted into metabolites. These
metabolites, if they prove mutagenic, will damage the DNA of
the histidine deficient *Salmonella* and cause some of them to re-
vert back to their original "wild" state. Once again able to
manufacture histidine, the bacteria grow into visible colonies
whose number provide a quantitative estimate of the mutagenic
potency of the suspected carcinogen. (McCann,J.,1976).

There are other fast, short-term tests for mutagenicity
(under these conditions always related to carcinogenicity)
which can be used to screen potential carcinogens and which do
not involve animal suffering. A reliable one, known as the Cell
Transformation test, uses a culture of neonatal Syrian hamster
kidney fibroblasts. Interestingly, the LD/50 is calculated as
part of the procedure, but here the "lethal dose" which kills
50% only slaughters cells, not whole animals as in its cruel
application in toxicology. (Styles,J.,1977).

Molecular Structure as Supporting Evidence in the Identification of Carcinogens

Since we live in an age increasingly dependent on chemicals in innumerable forms to sustain our way of life, and since
we have become wary of them as potential carcinogens, a mandate
to test has been handed by Congress to the federal agencies concerned. The National Cancer Institute's survey reports 7,000
chemicals tested by long-term animal bioassay up to 1975, of
which from 600 to 1,000 are potential carcinogens. But many of
these were selected for testing because they have a molecular
structure similar to known carcinogens; the number of chemicals
actually producing cancer in animals was estimated to be 221 in
an international survey (1978), and only 26 chemical substances
or processes have been implicated in cancer induction in humans
(6 by animal tests and 20 from epidemiological evidence). What

is suspected but not yet fully explored is the nature of so-
called promoting or potentiating agents which are themselves
not carcinogenic but which contribute to the development of cancer in subjects who have been exposed to the actual carcinogens.
(IRLG,1979,p.12).

Priorities of testing in this vast area must be established, and one method is by selecting compounds whose molecular structure, as mentioned above, resembles that of know carcinogens. (*Ibid*,,p.67). For instance, the polynuclear aromatic
systems, and N-nitroso groups, are arrangements of atoms known
to have carcinogenic properties. The more this kind of information can be developed through molecular analysis, the less
will empirical screening of thousands of chemicals via animals
be necessary - at least in the preliminary identification of
potential carcinogens.

Epidemiology

Very suggestive data on carcinogens often come from observations described as "epidemiological." A well-known finding
is that the incidence of cancer of the lung in women has risen
proportionate to the increase in cigarette smoking in that
group. Another is the observation of cancer incidence and mortality difference in people on a geographical basis - often
confirmed by an altered incidence in such a population after
migration. And clinical case reports may give early warning of
a potential carcinogen. (Nat'l. Cancer Advisory Bd.,1976,p.4).

But the incidence of cancer in workers exposed to chemicals,
or toxic reactions, is particularly diagnostic, and should be
continually monitored. This monitoring should not be confined
to the workplace; it should be continued after workers have re-
tired and, whenever feasible, studied in necropsy material after
their death. This would throw light not merely on mortality and

the more obvious morbidity, but also on the effects of low-level
exposure and on time trends - some cancers, for instance, may
not become manifest for several decades. (Anon.,1973).

The problem is how to instigate such studies, which would
both spare animals and benefit humans. Industry may well have
an ambivalent attitude toward epidemiological investigations.
While it should be obvious that such studies are the best way to
bring the hazards to workers and to those living near the factory
into focus, and to suggest preventive measures, there has been
fear that disclosure of toxic effects may lead to damage suits
and crippling injunctions. Even the government has indulged in
a cover-up of leukemia morbidity from nuclear exposure.

When monitoring has been attempted, it has probably relied
on animal toxicity testing, with LD/50, behavioral studies and
the like. The trouble with animal testing, aside from the
inhumanity, and the difficulty in comparing animals and man, is
that animals are subject to variables. For instance, as K.Z.
Morgan points out, sensitivity to leukemia induction [from
radiation], and life expectation, varies greatly with different kinds of mice. On the other hand, if one attempts to reduce the variables by developing inbred, carefully controlled
strains, such animals are even less comparable to the human,
who is a "wild or heterogeneous animal living in many types of
environment with various eating and drug habits, with many diseases and eccentricities, of various ages and so on." (Morgan,
K.,1979).

These variables in humans explain why one man may be vulnerable to cancer and another not, but the epidemiological
approach, although it can predict how many per thousand are
likely to develop cancer following exposure to certain known
carcinogens, cannot readily pick out the individuals who will
succumb to the disease. It works by hindsight, but what is also

needed are tests that can be applied to individuals to detect
early changes in the genetic material - DNA and RNA - of cells.

Chromosome Breakage: a Valid, Predictive Test?

It has been suggested that a test which does correlate with
threatened cancer is a high chromosome breakage rate. This has
been found in workers exposed to benzene, a carcinogen and a
highly toxic chemical. However, critics of some recent work
along these lines have pointed out that variability, the ghost
which haunts so many predictive tests, is also present here to
a degree which makes the forecast of cancer in any individual
very uncertain. Variables which are said to affect chromosome
breakage, according to Dr. J.R. Venable, director of biochemical
research for Dow Chemical Co., are factors such as "the viruses
workers might have had, medication they might have been using,
laboratory test techniques and even the season of the year."
The clincher for carcinogenicity is, of course, the clinical
appearance of the disease, but the fact that this may not show
itself for many years has so far prevented confirmation, at
least in the Dow study, of the benzene-chromosome breakage-
cancer linkage. (Severo,R.,1980).

When a single test is faulted because of variability, lack
of controls, or abbreviated follow-up, it simply means that a
broader spectrum of tests has to be administered if a "metabolic profile" known to be characteristic of the cancer-prone
is to emerge.

Hemoglobin Alkylation Test

One such test has recently been reported by Lars Ehrenburg
and his colleagues at the University of Stockholm. It bypasses
animal testing and requires merely a small sample of blood from
the person exposed to the toxic substance. It has been found
that the amino acids in the hemoglobin of the blood, cysteine

and histidine, react with vinylchloride and other alkylating
agents suspected of inducing cancer, at a rate which can be
simply related to that in DNA uptake, so this uptake in turn can
be easily calculated. Since hemoglobin cells have·an average
life of about four months in the blood stream, a sample records
exposure over about two months. A report states:

> "So far statistics are too few to determine whether the
> method can predict cancers in people exposed to alkylating
> agents. But it can pick out those who have accidentally re-
> ceived larger than normal doses. As one of the fundamentals of
> cancer treatment is to catch the disease early, high hemoglobin
> alkylation could be a valuable indication that the individual
> should undergo further observation." (Anon.,1979e).

Cancer Profiles by Mass Spectrometry

Another technique which shows increasing promise of being
able to sort out the numerous variables which occur in the
metabolic profile of an individual employs the mass spec-
trometer. An instrument that can identify and quantify some
150 components of urine alone may be extremely useful in "pro-
filing" persons with different types and stages of cancer and
identifying the metabolic abnormalities characteristic of their
disease. Besides urine, both blood serum and spinal fluid can
be analyzed with no harm to the patient. Epidemiological studies by this method are still in the future, but may eventually
be done on those who work with toxic materials. This kind of
investigation also bypasses animals completely, and can be performed on humans with no ethical objection. (Anon.,1978a).

To repeat: more accurate cancer prediction is not achieved
by testing a greater variety of animals but by exposing humans
to a greater variety of tests. To borrow an analogy from navi-
gation, by taking only one bearing you cannot determine where
you are at sea. But if you take bearings on two or more objects, and note where the lines cross, you have your position.

Five

IMMUNOLOGY

"Anaphylactic Shock Stress Without
Benefit of Anesthesia"

The immune system is one of the most complex in the vertebrate organism, yet everyone knows what it is to be allergic
to something, and we are familiar with what happens to sensitized people whose skin reacts with large swellings and wheals
to insect bites and poison ivy or to the child who wheezes and
may even strangle in the grip of asthma. A foreign protein,
the "antigen," enters the body, binds to a receptor on the surface of a lymphocyte (one of the white blood cells), and together they cause a reaction which may be anything from completely innocuous to immediately fatal "anaphylactic shock."
Medicines, such as the antihistamines, have been developed to
deal with these acute reactions. Testing is necessary in their
manufacture and this is fairly simple, although unfortunately
animals like guinea pigs are generally used to investigate
the possibility of anaphylactic shock. Thus in their Annual
Report under the Animal Welfare Act for 1976, Bristol Laboratories of Syracuse, New York, reported that 500 guinea pigs were
exposed to "anaphylactic shock stress without benefit of anesthesia" because in testing the drugs the "physiological state
of the animal cannot be compromised by prior use of sedatives."
(USDA/APHIS,1976,Bristol Labs.).

In Vitro Alternatives?

While it is true that the symptoms of anaphylaxis are all

too plainly visible in the whole animal, might not the reaction
be observed at a cellular or molecular level equally well, and
directly on human cells for which the drug under investigation
is ultimately intended?   J. Humphrey indicates just this when
he points out that principles worked out on experimental animals
are often inapplicable to man but "fortunately the great improve-
ments in tissue culture and micromethods have made it possible
to do many of the tests used on mice, etc., with human peripheral blood lymphocytes or easily obtained lymphoid tissue such
as tonsils." (Humphrey,J.,1978,p.109).

Demonstrating Anaphylactic Death in the Classroom - and the Alternative

In the Annual Report for 1976 under the Animal Welfare Act
submitted by the Dept. of Pathology, School of Dental Medicine,
University of Pennsylvania, the following justification is given
for subjecting animals to pain or distress without using pain-
relieving drugs:

"For classroom demonstrations guinea pigs are sensitized
for systemic anaphylaxis and then challenged *i.v.*[intravenously]
with antigen. Death in acute shock results within 5-10 minutes.
This demonstration is used to impress students with need to avoid
systemic allergic reactions in patients." (USDA/APHIS,1976,Penn-
sylvania, Univ.of).

Similar demonstrations were reported from Lehigh University,
Pennsylvania; Cleveland State University, Ohio; and Emporia
Kansas State College. Chicago College of Osteopathy, using 12
guinea pigs and 12 rabbits, gave as an explanation: "to the best
of our information, no comparable procedure available in an *in
vitro* system."

However, instead of witnessing the agonizing "sacrifices"
of live animals, students could have been impressed by a similar
demonstration in a film entitled, "ANAPHYLAXIS IN GUINEA PIGS,"
a 1960 production of a university bacteriology department. A

guinea pig sensitized to egg albumen receives a second injection
of the protein and is shown undergoing paralysis, bronchial constriction and, finally, death by suffocation. While a question
might be raised over the morality of subjecting even one guinea
pig to such suffering for educational purposes, the fact that
the film now exists offers a teaching alternative which could
spare the lives of many other animals. (California, Univ.of,
1960).

Defending the Body: Cell Eaters, Antibody Manufac-
turers, Helper and Killer Cells

The functions of the cells which constitute the immune system of the body, and names which have been given them, are stimulating to the imagination. Foreign substances entering the body,
the antigens, are seized on by the phagocytes, or "cell eaters."
These inhabit the lymphoid tissue, especially the lymph nodes and
the spleen, in whose cavities lurk the macrophages, or "big eaters." These cells prepare the antigens for presentation to the
next line of defense, the lymphocytes, which loiter along the
vessels of the lymphoid area or are swept through the blood
vessels in a jostling throng of red blood cells.

Two thousand billion of these lymphocytes are present in
the human body; every second about a million new lymphocytes
arise in the bone marrow and the same number die. Some of these
pass through the thymus gland and are known as T-cells, an equal
number are not thymus related and are called B-cells. Stimulated by the presence of antigens, the latter produce antibodies,
the defense protein molecules which at one end are so constructed that they can interlock with matching features on the surface of the antigens. This starts a militant operation to get
rid of the "foreigners" - chiefly bacteria and viruses.

The T-cells also defend against invaders, and are

particularly effective against fungi, parasites, cancer cells
and foreign tissue. There are different types of T-cells with
specialized functions, and these also proliferate on the arrival
of antigens. Among them are "killer" cells which attack and
destroy the invaders directly, and "helper" cells which assist
in B-cell differentiation and proliferation.

The next section introduces some immunological research
which has had fruitful results. The protagonists are an odd
couple: a sheep and a mouse.

Sheep's Red Blood Cells: a Research Tool in Immunology

When sheep's red blood cells are injected into a mouse,
they act as an antigen. B-and T-lymphocytes cooperate to produce antibodies which bind the sheep cells, against which the
mouse has now become immunized. However, it would be difficult
or impossible to repeat the experiment and come up with the
identical antigen-antibody complex. Numerous different antibodies, providing they and the antigen have at least some parts
(or "determinants") in common, may bind to the sheep's cells.
Also, each individual mouse produces a different antigen-antibod
combination, both in quality and quantity. Thus if one wanted
to standardize the antigen-antibody complex, i.e., the antiserum
so that it could be given to any and all mice to immunize them
against sheep's red blood cells, or if one wanted to isolate a
specific antibody for analysis, it would be necessary to use a
technique which would eliminate the differences between animals
and the variability due to the mixture of antibodies reacting
with the antigen.

Of course, neither mouse nor man in nature is going to need
an antiserum against sheep's red blood cells. The latter is jus
a convenient biological protein for research purposes. But, as

will be seen presently, a procedure worked out with these cells
can be applied to other antigens of more concern to humans - for
instance, cancer or influenza virus, or "histocompatibility"
antigens which are the ones which have to be neutralized if a
tissue or organ graft is to "take."

In immunological work, the interests of humanitarians and
researchers converge in their common desire to eliminate animals
as much as possible from experiments. To the experimenters, it
is the individual differences mentioned above which frustrate
purification and reproduction of biological products, even though
injection of antigens in any normal animal inevitably produces
antibodies and some degree of immunity.

"Search and Destroy" - the Antibody and the Tumor Cell

An application of immunology in cancer research and therapy
using chimpanzees as the animal model illustrates both the advantage of these new techniques and the humanitarian problem they
can create. H. Seigler has developed an antiserum against the
highly malignant and often fatal cancer melanoma by injecting human tumor cells into a chimpanzee. These cells contain a "tumor
associated antigen" (TAA), and the ape injected with them develops
an antibody against a human melanoma antigen. Seigler writes:

"There has long been interest in using tumor specific antibody to concentrate antitumor agents on tumor cells. In doing
so, many of the harmful side effects of chemotherapeutic drugs
might be avoided....One recent study had rather dramatic success
....The antibody was conjugated with an akylating chemotherapeutic and administered to a melanoma patient with disseminated tumor. Following this treatment, the tumor systematically re-
gressed....We have utilized radio-labelled chimpanzee anti-
human melanoma antibody in an effort to distinguish the specificity of the antibody and its ability to localize to the
harbored tumor. Both immunofluorescent studies and autoradio-
graphy studies of the removed tumor tissue have demonstrated that
the antibody did, indeed, fix specifically to the tumor cells."

Seigler has also found that the antibody is capable of
clinching a diagnosis of melanoma by specifically combining with
and thus identifying metastatic melanoma cells which the pathol-
ogist has not been able certainly to recognize. Once identified, the appropriate chemotherapy can be started. (Seigler,H.,
1977).

Remarkable as this experiment and its results appear, it
is unlikely that antibodies produced in another species could
be pure and specific enough to be generally successful in treating humans. Also, chimpanzees should not be exposed to highly
malignant tumors. A method was necessary which 1. would remove
the procedure from the actual animal to cells in culture, 2.
would permit the production of an antibody over a long enough
period for the cells expressing it to be isolated ("cloned"),
and 3. would result in its purification and its being made ab-
solutely specific for the antigen against which it was to act.

Monoclonal Antibodies "Immortalized" by Cancer

A very ingenious and "elegant" experiment which has accomplished just this was developed in 1975 by C. Milstein and G.
Köhler at the Medical Research Laboratory in Cambridge, England.
It is known that a B-cell-derived malignant lymphoid cancer (myeloma) of mice will grow indefinitely in the laboratory. (Köhler
G.,1975). Cells from this if placed in a culture medium containing polyethelene glycol will fuse with normal B-lymphocytes from
the spleen or lymph nodes of a mouse. This mouse has previously
been injected with sheep's red blood cells and therefore the
B-lymphocytes are producing antibodies against the sheep cells.
The fused cells have immortality conferred on them by the continuously self-replicating cancerous part. The culture, however
contains not only the fused cells but also unfused myeloma and
spleen cells. It is possible to get rid of the unfused cells,

leaving only the desired fused or "hybridoma" cells, by juggling
the chemicals in the broth medium in which the cells are grown
so that it fails to nourish first the unfused myeloma cells
then the unfused spleen cells. These die out, leaving the hybrids energetically proliferating and secreting antibodies
against the sheep's blood cells. (Staehelin,T.,1978,p.133).

At this point there is still a mixture of antibodies in
the cell culture. Using special techniques such as autoradio-
graphy and isoelectric focusing it is possible to identify sub-
populations or "clones" of B-cells, and to segregate the one
among these which produces the specific antibody required. It
is called "monoclonal" because all the cells of a clone are de-
scended from one cell and make antibody molecules with identical
connectors at one end which bind to a specific antigen. Even
then, the culture has to be constantly monitored to prevent
overgrowth by variant cells appearing through spontaneous mutations, but repeated subcloning will segregate the cells of
desired specificity from these mutants. Antibodies from these
are far purer than any that could be obtained from the serum of
rabbits or other animals which had previously been injected with
an antigen. Furthermore, the hybridomas can be frozen and
preserved for future use.

Cell Culture as an Alternative

One method of assuring a stable and long-term source of the
monoclonal antibody is by injecting these hybrid clones into mice
to produce tumors from which the antiserum can be harvested. A
humane alternative is not to use the whole animal but to continue the growth in cell culture and thence to collect the antibody. There is also the advantage of eliminating the contamination of whatever other antibodies there may be in normal mouse
serum. There are many exciting applications of this research;

in fact the possibility of making antibody molecules which will unerringly seek out and detect specific antigens has been called a revolution in immunology. It is not the simple antigen-antibody reaction such as occurs when an antigenic pollen hits the antibodies on the mucous membrane of the nose of a hay fever sufferer and produces the familiar symptoms. In that case there is likely to be a mass of different antibodies responding to a mixture of antigens. But for diagnostic purposes, for developing a medicine specifically designed for a particular disease germ or cancer (and delivering it to a certain location and nowhere else), for drawing fine distinctions between cellular and subcellular structures, this new technique is proving an exquisitely sensitive research and therapeutic tool.

The Next Step: of Mice and Men

Dr. T. Staehelin, asked at the Roche Research Foundation Symposium in Basle (1977) to discuss the sensitivity of mouse antibodies to human antigens, replied that the hope of the future lay in using human lymphocytes in an *in vitro* (test tube) immune response.

"This approach may eventually turn out to be a necessity and may have a great advantage....If it is possible to fuse human lymphocytes with an established human lymphoblastoid or plasmacytoma cell line, then we may be able to produce monoclonal human antibodies. Of couse the potential here would not only be diagnostic, but eventually even therapeutic." (*Ibid.* p.139).

Köhler and Milstein in the experiment described above fused mouse lymphocytes with a mouse myeloma cell line. Staehelin's 1977 prediction referring to the possibility of using human cell to produce antibodies from hybrids has already come to pass in this rapidly expanding field of research. H. Koprowski and others (Koprowski,H.,1978,p.17) took lymphocytes from the cerebrospinal fluid of a patient with a herpes zoster encephalitis.

These were hybridized with mouse myeloma cells and were found to segregate into clones that produced human antibodies and clones that did not.

By the usual purification techniques it should be possible to select those clones whose antibodies are specific to the encephalitis antigens, with a potential of diagnostic, therapeutic or research use.

And Finally, the All-Human Hybridoma

Most of the hybridoma research reported up to the present has used mouse myeloma cell culture as the "immortal" part of the antibody producing system. But human lymphocytes can also be transformed into cancer and, hence, into immortal cell lines, by exposure in the laboratory to a virus called Epstein-Barr. This creates a "lymphoblastoma," and there seems no reason why this could not be linked with normal human lymphocytes producing specific antibodies. If the donor of the latter were suffering from a cancer, such as melanoma, one would expect the antibody to be specific for a melanoma tumor virus; if the individual had been immunized with tetanus toxoid, one would look for a tetanus antibody. Even without the intervention of animals, therefore, a hybridoma capable of producing antibodies of exquisite specificity to any known antigen can be made.

F. Melchers and his co-editors of "Lymphocyte Hybridomas" conclude the Preface with this paragraph:

"Monospecific antibodies have been the dream of many immunologists for long....Lymphocyte hybridomas have made the dream become a reality. It is, therefore, predictable that lymphocyte hybridoma cultures are likely to replace many rabbits in many research laboratories." (Melchers,F.,1978,p.XVII).

A Laboratory Without Animals: the Patient's Body

But the last word on this subject has not by any means been said, so varied are the possibilities of immunological research.

The investigations reported above were chiefly concerned with developing an antibody, or antiserum, which might ultimately be injected into a patient to combat a cancer-bearing tumor associated antigen (TAA) known to be the specific target for that antibody. S. Leong and associates have taken cells from a melanoma sufferer, combined them with an immunological strengthener or adjuvant known as BCG, and injected them, not into a chimpanzee, as Seigler did in the experiment described above, but back into the patient himself. This produces an antibody against his particular TAA, although one too weak to arrest the disease. Nevertheless, this antibody can be a useful tool. Leong and associates have also been working with cultures of human melanoma cells and have isolated various tumor associated antigens from the membrane of these cells. If the patient's antibody is extracted and combined with these antigens, it can pick out the one which is identical with the patient's, and this can be purified and developed into a vaccine. This in turn can be injected into the patient and may be able to mobilize enough antiserum in his blood to destroy the cancer. (Leong,S.,1978).

Epstein-Barr Virus to the Rescue?

Of course it is not possible to risk injecting live or even inactivated cancer cells as a vaccine to produce antibody against cancer into anyone other than a patient who already has the disease. It is possible, as we have seen, to grow specific antibodies in culture outside the body, but this is a roundabout way to produce them. It's much more effective if the body acts as its own laboratory and makes antibodies just as it does in response to a shot of tetanus vaccine. In the case of cancer, the solution may come through one of those tumor associated antigens on the membrane of the cancer cell.

The Epstein-Barr virus, which causes relatively innocuous mononucleosis - the "kissing disease" of the young - is also responsible for a lymphoid cancer, or lymphoma. (Its ability to transform laboratory cultures of human lymphocytes into cancer was mentioned above). Furthermore, a tumor associated antigen is a normal inhabitant both on the surface coat of the virus and on the lymphoma cell membrane, although this antigen itself is neither infectious nor cancer-producing. However, if injected into an animal, it does produce antibodies and renders the animal immune to mononucleosis. If animals, why not humans? and this may lead to protection against the lymphoma too - which especially affects African children - and against Epstein-Barr nasopharyngeal cancer, a distressing affliction of Chinese men. (Anon.,1980d).

Neither this vaccine, nor the anti-melanoma antiserum developed through hybridization is yet available (in early 1980) for therapy. But should cancer actually turn out to be an immunological aberration, we are surely "getting warm" in this scientific game of hide-and-seek. And if, as seems possible, anticancer viruses of the future can be grown on human cell cultures, rather than in animal models, then the rabbits, and monkeys and ourselves, for different reasons, have much to look forward to.

Six

INHALATION OF TOXIC SUBSTANCES

Development of Bronchodilator Drugs

As mentioned on p.239, Warner-Lambert Research Institute reported that in 1973, for bronchodilator drug screening, approximately 1800 guinea pigs were subjected to the distress of an aerosol of histamine. Their report to the USDA adds: "If the animals are not protected by a bronchodilator they develop intense bronchospasm and will ultimately expire of asphyxia. These studies must be done in the unanesthetized state in order to determine the endpoint." This experiment is apparently routine at Warner-Lambert since it occurred again in their report for 1976 under the Animal Welfare Act (cf.p.241). In the latter year no less than 3,328 guinea pigs were reported as suffering "pain or distress" in similar tests "for screening and development of new drugs." (USDA/APHIS,1973 & 1976,Warner-Lambert Res.Inst.).

Alternatives: Organ Cultures

Since the drugs concerned have a rapid and direct action on the respiratory system, the situation seems one in which organ culture could replace or at least reduce the use of animals. Organ cultures of rings of rat trachea (windpipe) have been maintained for four to six weeks, and chick, hamster and even human lung tissue has been cultured.

Two experiments using trachea cultures from guinea pigs may be cited. One compared the contrasting effect of

104

spasm-producing histamine and the bronchodilator isoproterenol in live animals and in tissue culture. (Douglas,J.,1977).

The second experiment, from the Harvard School of Public Health, compared the effects of histamine and another spasmogenic agent, acetylcholine, on trachea and lung cultures. The different pharmacological response to these two drugs observed in the cultures was consistent with what had been noted previously in animals. (Drazen,J.,1978).

Furthermore, since it is necessary to know the effects of the drugs on systems of the body other than the lungs, organ cultures can be made, for example, of heart muscle, (Laity,J., 1971), or even of the human vein (Gailer,K.,1971).

Alternative: Human Volunteers

Once the basic safety of drugs has been established, an early use of human volunteers is valuable both in drug development and in clinical testing. It has the great advantage of sidestepping the confusion that often occurs with drug testing in species other than man.

G. Marlin and P. Turner at St. Bartholomew's Hospital, London, investigated the effects of a new drug, rimiterol hydrobromide, potentially useful in the treatment of severe asthma, on seven human subjects who were sufferers from various allergies and hay fever, or were smokers, and who developed bronchial constriction on histamine inhalation. The tests were similar to the Warner-Lambert experiments on rats and showed the protection afforded by rimiterol (and two other comparable drugs) against histamine-induced bronchoconstriction. (Marlin,G.,1975).

Asbestos

Asbestos fibers have been found to produce lung cancer, both in workers exposed to the dust and experimentally in rats,

either by inhalation or by inoculation of the fibers into the
lungs.

Alternatives

Since these experimental methods are time-consuming (results cannot be expected in less than three years from the
initiation of the investigation), Rajan and co-workers obtained
human pleura from a lung operation and exposed it to asbestos.
Several of the organ cultures lived for up to 8 days, and the
asbestos caused marked overgrowth of mesothelial cells, some
of which began to invade the underlying tissue (in cancer-like
fashion). The researchers comment: "Organ culture systems
should be useful in investigating the effect of various fibres,
chemicals and perhaps carcinogens in a relatively short period."
(Rajan,T.,1972).

Andrew Sincock, associated with the same lung research unit
in a mining district of South Wales as the above investigators,
and his co-worker M. Seabright, carried the work further by exposing cultures of Chinese hamster cells to asbestos fibres.
They found that

"multiple abnormalities at both the chromatid and chromosomal levels were produced in Chinese hamster cells....It is
envisaged that primary cells will be cloned for specific chromosomal abnormalities and then introduced into laboratory animals
to ascertain whether or not they induce tumors in vivo."
(Sincock,A.,1975).

Although these investigators plan to study the cancer potential of these abnormal cells by injecting them into animals,
P. Noguchi and associates at the Bureau of Biologics, U.S. Food
and Drug Administration, have shown that chick embryonic skin
in organ culture can serve as well as the living mouse does for
the identification of cancer cells. The work of two of these

associates, J.C. Petricciani and R. O'Donnell, was partly supported by the Lord Dowding Fund (London). (Noguchi,P.,1978).

Tobacco and Smoking

Dog-smoking experiments received much publicity a decade
ago when Drs. Auerbach and E. Hammond, who forced 86 beagles
to inhale the combustion products of 415,000 cigarettes, sometimes for as long as 2 1/2 years, announced that 12 of the dogs
had developed lung cancer. (Altman,L.,1970,p.1).

Human smokers in never-ending supply demonstrate the same
results, but, for the animals, there is no let-up. The grotesque experiments keep pace with the relentless promotional
advertising of the cigarette manufacturers, now chiefly aimed
at the young. The Louisville Courier-Journal, June 18, 1978,
(Russell,J.,1978) describes a study under Dr. Lester Bryant at
the University of Kentucky Tobacco and Health Research Institute
which uses two dozen or so stump-tailed monkeys. Strapped into
chair, the monkey wears a mask which covers the nostrils while
a bit holds the mouth partly open. A machine delivers smoke at
40-to 50-second intervals and the animal must inhale the smoke
if it is to breathe at all. Some of these former inhabitants
of the jungles of Vietnam and Thailand are forced to smoke from
one to three packs of cigarettes a day.

Alternatives

An alternative series of experiments, not inflicted on animals and more pertinent to humans, have been described by C.
Leuchtenberger and associates. They exposed human lung cultures
to marijuana and tobacco smoke in a smoking machine and demonstrated abnormal cell growth and chromosome disturbances, pre-
cursors of cancer. (Leuchtenberger,C.,1973).

Another link between smoking and cancer has been demonstrated in humans - without resorting to animal experiments -
by E. Yamasaki and B. Ames. They concentrated the urine of
21 nonsmokers, and that of 7 smokers who inhaled and smoked
ordinary (non low-tar) cigarettes. They subjected the concentrates to a resin adsorption technique, then to an assay
by the *Salmonella* bacteria test for the presence of mutagens
(pre-cancer substances). They found the following: none of
the nonsmokers had mutagens in their urine; all of the smokers
did have them. Four more smokers were tested: two of these who
did not inhale had no mutagens in the urine, and of the other
two, who inhaled but smoked low-tar cigarettes, one had no
mutagens in the urine but the other did. (Yamasaki,E.,1977).

For the record, it should be noted that although smoking
relatively low-tar and nicotine cigarettes as against ordinary
ones results in slightly lower death rates, those rates are
still far higher than the mortality of people who never smoked
regularly. (Am.Cancer Soc.,1979).

Chromates

Hexavalent chromium compounds, particularly chromates,
are occupational hazards to those employed in the chrome-
plating and chromate-producing industries. They have been implicated, through the inhalation of the dust and vapor, in diseases of the respiratory tract, including bronchopneumonia,
chronic bronchitis and tracheitis, and are thought to be responsible for the high prevalence of bronchiogenic cancer among
chromium workers. (NAS Com.on Medical...,1974,p.42).

Alternatives *In Vitro*

Although studies of chromate toxicity have been made in
the whole animal, an increasing number are being done by means

of organ culture, since thin slices of trachea will remain
alive in suitable media - *in vitro* - for as long as six weeks.
M. Mass and B. Lane at the State University of New York (Stony
Brook), removed tracheas from rats in a non-recovery operation
and studied the effect of chromates both on the freshly excised
tissue and on tracheal rings grown in culture for up to two
weeks. They tested different concentrations of the metal, and
observed through the microscope that cells were dying at concentrations of chromate in the 10 - 100 microgram/milliliter
range, far below the concentration which stopped the ciliary
motion of mucus. They concluded from this and other evidence
that the chromates did not act primarily as "ciliostats" but
that the chemical injury might be occurring at the cell membrane. (Mass,M.,1976).

Radionuclides

The proliferation of nuclear power systems has sparked
numerous experiments subjecting animals such as beagle dogs and
hamsters to the inhalation of radionuclides. For a discussion
of these experiments, see p.138, Chap.VIII, "Radiation."

Seven

THE PLIGHT OF THE NONHUMAN PRIMATE

Getting Ready for Their Questions

Up to this point each chapter has had a major area of
research or testing for its theme, but here the animals - nonhuman primates, or primates for short - are themselves the subject. While it might seem invidious to pick animals with the
closest evolutionary relationship to humans, rather than the dog,
"man's best friend," or the whales and dolphins, aspects of
whose intelligence appear to rival our own, a number of circumstances add up to make the plight of these animals altogether
special. They are endangered as a species; their activity, derived from an arboreal lifestyle, condemns them to distressing
physical restraint in the laboratory; they *are* highly intelli-
gent and sensitive; and probably the many experiments they are
subjected to in the U.S. have been statistically surveyed more
than that of any other species. Finally, as a result of the
recent breakthrough in teaching chimpanzees like Washoe, Lana,
Lucy and Sarah, and the gorilla Koko, to communicate through
sign language and other symbolic methods (Lana uses a type-
writer), in the not too distant future we may hear some comments
from one of these apes, perhaps not altogether complimentary,
on human behavior. (Bourne,G.,1977 and Hayes,H.,1977).

Because of these special factors, it is worth looking now
at the plight of this one species as a whole, although numerous

mentions of other experiments on primates will be found throughout the book and listed in the Index. This chapter may cause
the reader to ask why we presume to treat our fellow-primates
so abominably. Better ask ourselves before the apes do....

Seven Monkeys in Six Cubic Feet

In June 1974 I visited the Animal Center Section, National
Institutes of Health, Poolesville, Virginia. Dr. Albert E. New,
then Head of the Primate Quarantine Unit there, let me observe
the arrival of cages from India filled with rhesus monkeys -
each cage only 2 1/2 ft. long x 1 3/4 ft. wide x 1 1/2 ft. high,
crammed with seven animals weighing 4 1/2 to 6 1/2 lbs. apiece.
They had been flown via London, where the small food and water
containers in their cages had been re-filled; nevertheless,
they were very thirsty when they arrived by truck at Poolesville from the Washington airport. At the Center they had to be
quarantined and treated for diseases, principally tuberculosis
and shigellosis, which they were likely to have brought with
them - diseases contracted from humans in their native land.

Number of Primates Used in Research

Five to six thousand monkeys a year were being received at
Poolesville at the time of my visit, and 75,000 were being imported into the U.S. in 1973 for all purposes. Of these, about
55,000 were for biomedical use. The remainder were for exhibition and the pet trade, but the total included an estimated
10,000 which died during quarantine and 'conditioning.' The
number of primates maintained in the U.S. institutions in 1973
was estimated at 67,900.

During the development of polio vaccine in the late 1950's,
the U.S. was importing rhesus macaques, alone, at the rate of
200,000 a year. During the years 1958-1976 over two million

were imported. As a result of this high level of commercial
exploitation, plus deforestation in the countries of origin,
as well as expanding agriculture, urbanization and market hunt-
ing for meat, world primate populations were rapidly declining
in the 1970's.

Concerned about the increasing scarcity of primates for
biomedical research and pharmaceutical production, the National
Institutes of Health sponsored a study by the Institute of
Laboratory Animal Resources, completed in 1975, on the current
use and availability of these animals both in the U.S. and
abroad. (ILAR,1975).

To Procure 65,000 Primates for Research, another 85,000 are Sacrificed

This ILAR survey, from which most of the statistics in the
above paragraphs are taken, reports that losses of primates as
a result of collecting methods in countries of origin "have
been estimated by some to be as high as 50% of the numbers
captured." In 1967, 62% of vervet monkeys were estimated to
have died during collection in Kenya from a variety of factors,
including death "at the collecting station, death in unattended
traps, young released without their mothers, and the culling
rate by dealers who accept only animals of a specific size."
The World Federation for the Protection of Animals reports that
in "Colombia and Peru large areas are completely cleared of
monkeys. Indians in remote areas are now commercial hunters
and trappers. Four or five monkeys of even robust species die
for every one successfully exported." About 750 chimpanzees ann
ally are required by the world's laboratories but "for every
chimpanzee which reaches a laboratory six or seven others die
during trapping, holding by dealers, transport and quarantine.

This brings the total to about 5,000...which are sacrificed to
science *each year*." (Harrison,B.,1971).

The most conservative of the above estimates indicates
that to produce approximately 65,000, net, primate imports
used by the U.S. in 1973, another 85,000 animals, including
the 10,000·which die in quarantine or in conditioning after
reaching this country, were destroyed.

In Nov. 1976 international action resulted in all primates
being placed either in Appendix I (species threatened with extinction) or Appendix II (species which may become threatened
and for which regulation of trade is, therefore, necessary) of
the Convention on International Trade in Endangered Species.
This requires the exporting country, in every instance, to
issue a certificate that shipment of the animals will not be
detrimental to the survival of the species in the wild. (Smith,
R.,1977).

Because of the diseases they transmit, but perhaps also in
the hope of increasing the number available for research, the
U.S. Dept. of Health, Education and Welfare several years ago
prohibited the importation of primates except for scientific,
educational or exhibition purposes. Nevertheless, even with
the pet market eliminated, the decline in imports continued.
In 1977 primate imports into the U.S. stood at 29,000, of which
roughly one-third originated in India and were probably rhesus
macaques. (Anon.,1978d,p.10).

India Bans the Export of Primates

In 1955 the government of India had banned the export of
rhesus monkeys as a result of fatalities during export. It was
persuaded to rescind the ban several months later, but at that
time entered into an agreement with the U.S. to insure humane
treatment for any primates which might be received from India

in the future. In the agreement India specifically prohibited
the use of the monkeys in atomic blast experiments and space
research, allowed their use in medical research and the production of antipoliomyelitis vaccine, but required that each
purchase order be certified to this effect. There followed
nearly two decades of rhesus imports numbering well over a
million animals, the time slipping by without much attention
being paid to what was happening to the monkeys. Then came
the development of the neutron bomb. Simulated radiation effects of this weapon had been tested on various animals since
1966 at the Armed Forces Radiobiology Research Institute(AFRRI)
in Bethesda, Maryland, but it turned out, and was reported in
the *Guardian* (Tucker,A.,1977), and the *Washington Post* (Anon.,
1977b), that AFRRI had been using rhesus monkeys in these and
other experiments - in fact 1379 primates were known to have
been so used from 1972-1977.

The effects produced in these sensitive creatures by massive doses of radiation included diarrhea, vomiting (sometimes
up to as much as 50 times an hour) and tremor, while their
rapidly declining ability to function was tested on a treadmill
on which they had been trained to run, or rather forced to run,
in order to avoid punishing electric shocks. When they became
too incapacitated to turn the treadmill, they were shocked, and
if a "supralethal exposure" to radiation had been administered,
this torture continued until death. (McGreal,S.,1978).

Representatives of the International Primate Protection
League, as well as British humanitarians Muriel, Lady Dowding
and Jon Evans, brought this abuse of Indian rhesus to the attention of the Indian Prime Minister, Shri Morarji Desai. Other
apparent violations of the 1955 agreement were also publicized
in India, so, following an investigation of the whole matter of

monkey export, the Prime Minister decided to ban the trade.
This became effective on Apr. 1, 1978.

Recently, Shirley McGreal, Co-Chairwoman of IPPL, received
new assurance on the ban from Prime Minister Gandhi. On
Mar. 8, 1980, Mrs. Gandhi wrote:

"Dear Dr. McGreal, I have seen your letter of the 21st
February about protection of Indian monkeys. We shall certainly do what we can to minimize cruelty to animals and also to
humans. The ban on export of all types of monkeys from India
continues and there is no proposal to reopen this now...."
(McGreal,S.,1980).

The Big Importers Carry On

These events have sent shock waves through the research
establishment. Hazleton Laboratories Corp., an international
biomedical research concern which along with Charles River
Breeding Laboratories is the largest importer of rhesus monkeys
into the U.S., had an operating profit of 30% on $1.2 million
sales of these animals in fiscal 1978, but this extremely profitable part of Hazleton's $30 million annual business nose-
dived after the ban: the company's research animal sales were
down 35% in the first nine months of 1979. However, Hazleton
and other importers turned quickly to Malaysia, Indonesia and
the Philippines for another source of primates, the cynomolgus
macaque (M. fascicularis), although a financial report on the
company laments that "the cynomolgus monkey is only marginally
profitable because of a high mortality rate in shipment and the
high price paid to exporters." (Roth,N.,1979,p.18).

However, despite the mortality, the importation of the cynomolgus, according to the U.S. Dept. of Commerce, was stepped
up from 9,500 in 1977 to 16,100 in 1978, thus partly compensa-
ting for the rhesus ban, although the rhesus is still required
by Food and Drug Administration regulations for vaccine testing.

Primate imports were around 30,000 for both these years, to whic
must be added those born in the U.S. at N.I.H. Primate Centers,
at Hazleton's Texas Primate Center, at Charles River Breeding
Laboratories' Key Lois in Florida, and elsewhere - perhaps
another 6,000.

Domestic Breeding of Primates

The Interagency Primate Steering Committee estimates that
"the present breeding colonies of rhesus in the U.S. should yie
a net of 5,500 annually by 1980." (IPSC,1977,p.B-2). Thus the
contention of the International Primate Protection League seems
valid: that the appeal by "primate procurement politicians" to
officials of foreign countries urging them to release monkeys
for desperately needed polio vaccine testing is spurious, becaus
domestic production of rhesus is now sufficient to meet that
need. (Anon.,1979d,p.5). The motive is more likely a preference
for the cheaper imported rhesus, which cost about $250 before th
Indian ban as against the up to $800 now charged for the captive
bred animal.

Including the rhesus mentioned above, the total U.S. feder
ally supported production of primates projected to 1980 is estimated at 9,540 - see *Table 1*. (IPSC,1977,p.D-1). There has
been some commercial breeding without government support, but
since this has been relatively insignificant, there are still
28,000 primates to be imported if the present level of experimentation is to be maintained. What are the major biomedical
uses of primates and how many of these are really necessary? T
question must be considered against the background of species
threatened with extinction throughout the world and the cruelty
and wastage of animal life inseparable from the trapping and
long-distance transportation of sensitive primates.

U.S. DOMESTIC PRODUCTION
OF NONHUMAN PRIMATES
SUPPORTED BY FEDERAL GOVERNMENT
BY SPONSOR AND SPECIES

1980 PROJECTED LEVELS

Sponsor	Rhesus	Cynomolgus	Other Macaque	Saguinus Marmoset	Common Marmoset	Owl Monkey	Chimpanzee	Gibbon	Baboon	African Green Monkey	Squirrel Monkey	Other	TOTAL
Federal Gov't.													
DHEW ADAMHA CDC	50												50
FDA	2000					30				50			2080
NIH Extra-mural[1]	3205	250	460	100	100	25	40		40	150	690	80	5140
Intra-mural	1100	30	10	360	20	20	20	10	20		110	380	2080
DoD Army						60							60
Navy	130												130
TOTAL	6485	280	470	460	120	105	90	10	60	200	800	460	9540

Table 1. (IPSC,1977,p.D-1)

Key to abbreviations: DHEW: Dept. of Health, Education and Welfare; ADAMHA: Alcohol, Drug Abuse and Mental Health Administration; CDC: Center for Disease Control; FDA: Food and Drug Administration; NIH: National Institutes of Health; DoD: Dept. of Defense.

[1] *Supplies ADAMHA with approximately 100 rhesus monkeys per year.*

Biomedical Use of Primates in the United States

The Interagency Primate Steering Committee identifies four main groups of activities in the biomedical use of primates, the 1977 requirement for which was estimated at 33,912 - see *Table 2* (*Ibid.*,p.A-4). I quote below from IPSC's *National Primate Plan.*

1. Legal and Regulatory

First biomedical use: "*Legal and Regulatory.* Procedures prescribed by law or regulation; e.g., those pertaining to production, potency testing, and safety testing of viral vaccine seed suspension and vaccine lots of licensed products."

The principle vaccine in this group is polio virus, the production and testing of which requires about 4,000 rhesus; approximately another 2,000 are required for the neurovirulence testing of other vaccines, including those for adenovirus, measles, mumps and rubella. Unless the vaccine is defective, the monkeys do not suffer since the injections (including the ones into the brain) are made under anesthesia and the animals are usually protected from the disease by the vaccine. The trouble here is that after a period of observation the monkeys have to be killed and their tissues examined to be sure no destruction of nerve tissue has occurred.

For a full discussion of the polio virus vaccine procedure and suggestions for reducing the numbers of animals used, see p.194 ff.

2. Biological Production

The second principal biomedical use of primates is described as follows: "*Biological Production.* Preparation of biological material such as experimental vaccines, tissue cultures, serum products, and biological diagnostic reagents. This

ESTIMATED 1977 REQUIREMENTS
FOR NONHUMAN PRIMATES
BY SPONSOR AND NATIONAL HEALTH NEED

Sponsor	Required by Law or Regulation	Production of Biologics	Test-ing	Re-search	TOTAL
Federal Gov't.					
DHEW					
ADAMHA				600	600
CDC		200	75	270	545
FDA	2,450		220	410	3,080
NIH					
Extramural				11,430	11,430
Intramural				1,355	1,355
DoD					
Air Force				781	781
Army			250	1,930	2,180
Navy				675	675
Other				560	560
EPA				770	770
ERDA				110	110
NASA				275	275
NSF				254	254
VA				862	862
Other Agencies					
Industry					
Pharmaceutical/ Biological	4,500	800	2,920	1,200	9,420
Other					
Other[1]				1,015	1,015
TOTAL	6,950	1,000	3,465	22,497	33,912

Table 2 (IPSC,1977,p.A-4)

Key to abbreviations: See *Table 1*, p.117; also EPA: Environmental
Protection Agency; ERDA: Energy Reserve and Development Adminis-
tration; NASA: National Aeronautics and Space Administration; NSF:
National Science Foundation; VA: Veterans Administration.

[1]
*Other includes requirements of activities funded by state and
local governments, foundations and academic institutions.*

includes activities using primates for production phases of
these materials which are not specifically required by law or
regulation." (*Ibid*.,p.A-2).

No more than 1,000 primates are used in this category:
800 by industry and 200 by the Center for Disease Control (among
the operating components of the Center are the Bureaus of Laboratories, Tropical Diseases and Smallpox Eradication).

The Interagency Primate Steering Committee points out that
primates were first used for these products when they were relatively abundant and cheap, in addition to being effective sources
for the required material. Since the first two reasons no longer
prevail, the Committee recommends that the users of primates re-
examine the need for these animals "in recurring and production
activities, ensuring that there is no acceptable alternative to
their use and, where necessary, place increased emphasis on the
development of new techniques and procedures in order to further
reduce this need." (*Ibid*.,p.12).

However, it should be added that the production of vaccines,
antitoxins and hormone products, and the taking of blood or serum
is not necessarily a painful experience for the animals concerned.
The same cannot be said for the potentially very distressing testing of the virus or toxin for virulence in control animals which
have not been protected by vaccine or antitoxin, see p.193-194.

3. Testing for Efficacy and Safety

The third biomedical use is "*Testing*. Evaluation of
efficiency or safety of products for prophylactic, therapeutic
and nutritional purposes. This includes activities concerned
with efficacy and safety testing of vaccines prior to licensure,
and of other compounds, materials, apparatuses, or other devices
prior to approval for marketing." (*Ibid*.,p.A-2).

In 1977, the distribution of primates for the above pur-
poses was as follows: Center for Disease Control, 75; Food and
Drug Administration, 220; Army, 250; Industry (Pharmaceutical/
Biological), 2,920; Total, 3,465.

When I visited the Food and Drug Administration's Bureau
of Radiological Health, Rockville, Maryland, in 1974, I saw a
number of these monkeys. Each was sitting on a wire floor in a
small cage. They looked bored and listless, except for one who
was jumping up and down and banging against the metal. They
were being tested for the effects of radiation exposure - for
instance, from a microwave oven. They had been trained to do
some repetitive work, and each day they were rolled in a chair
to the "workroom" to see whether their performance had been
affected by the radiation.

Among the 250 monkeys assigned to the Army could well have
been some of those who died in experiments at the Armed Forces
Radiobiology Research Institute in Bethesda, Maryland, after
exposure to up to 200 times the lethal dose of neutron radiation.
The investigation sought "to model the effects of prompt ex-
ternal ionizing radiation on the physical performance capability
of combat personnel...." As already mentioned, these monkeys
were trained by electric shocks to run on a treadmill and their
performance was evaluated before and after radiation. (Anon.,
1979a,p.4).

An idea of the industrial testing which needs nearly 3,000
primates can be had by studying the annual reports of research
facilities submitted to the U.S. Dept. of Agriculture under the
provisions of the Animal Welfare Act. A rough check of the reports for the year 1976 reveals that about 1,100 primates are
mentioned under the heading "Pain - no Drugs: number of animals

involving pain or distress without use of appropriate anesthetic, analgesic or tranquilizer." There are numerous examples of animals trained to push a lever to avoid an electric shock, then given a drug to see whether it speeds up or slows down the avoidance rate. An example of a nutritional test is one performed by Hoffmann-La Roche of Nutley, New Jersey, on a group of 144 primates; the absence of pain relief was explained as follows:

"Induction of dietary deficiency disease by feeding subminimal diet. Test compounds are administered to evaluate their therapeutic activity. Anesthetics, analgesics and tranquilizers may interfere with test compound activity and obscure clinical evaluation." (USDA/APHIS,1976,Hoffmann-La Roche).

The rest of the primates, including thousands used by pharmaceutical concerns, are enumerated in the annual reports under the Animal Welfare Act, but, since their pain was either relieved by drugs or was inconsequential, the nature of the experiments is not required by the regulations to be further identified. No doubt many of them were used in testing the "efficacy and safety of vaccines prior to licensure."

4. Research

The fourth major area of primate biomedical activity is *Research*. This is by far the largest area of use and in 1977 accounted for 22,497 animals. Instead of quoting the brief paragraph on the subject from the *National Primate Plan*, I turn to a more comprehensive description, in *Nonhuman Primates,* a publication based on a survey by the Institute for Laboratory Animal Resources on the usage and availability of primates for biomedical programs. The survey covered the year 1973.

This description begins with a mention of the three areas of use, vaccine production and testing, pharmacology, and toxicology, which we have already mentioned. These account for the first 38% of all primates.

"Studies of various diseases, including experimental surgery, account for the second largest demand, or 36 percent of
the total primates. Cancer studies consume marmosets over all
other species, including rhesus. Both marmosets and night monkeys are used in proportionately high numbers along with rhesus
in studies of infectious diseases. This probably reflects the
use of marmosets for work in hepatitis and night monkeys in
malaria. Nearly as many baboons as rhesus are used in experimental surgery, which is striking when one considers their
relative total numbers in use. Neurophysiological studies account for 16 percent and represent the second largest use for
each of 3 species: the traditionally available rhesus macaque,
the relatively large-brained squirrel monkey, and the night
monkey, which is favored for its large eyes and nocturnally
adapted retinas. The final 10 percent of total primates are
used in physiological and behavioral studies." (ILAR,1975,p.37).

Alternatives in Infectious Disease Research

Malaria

Malaria research has involved the use of the night or owl

monkey (Aotus trivirgatus), particularly the North Colombia

subspecies. They have become difficult to obtain because of

dwindling populations as a result of habitat destruction. How-

ever, in 1976 Trager and Jensen at Rockefeller University

succeeded in growing plasmodia, the parasites which cause mala-

ria, as a continuous culture in a suspension of human red blood

cells (Trager,W.,1976); it has since been possible to make the

parasites complete their life cycle in laboratory culture. This

should provide an alternative to the use of the monkey as a

model for the disease process. Research is continuing in the

hope of developing a vaccine from the gametocytes (sexual stage

of the parasites) which infect mosquitoes from human blood. This

could prevent the transmission of the disease to these insects.

A further development may produce a vaccine which will protect

an individual from the bite of an already infected mosquito.

Hepatitis A

The use of the rare marmoset monkey in hepatitis A research
has been mentioned. Up to now, only the livers of this diminish-
ing South American species have been suitable for culturing the
virus of infectious hepatitis. An alternative technique of
growing the virus - on cell cultures originally taken from the
kidney of a rhesus monkey fetus - has recently been announced by
P. Provost and M. Hilleman, at the Merck Institute for Therapeutic Research. Unfortunately, fetal rhesus monkey cells stop reproducing after 12 population doublings in culture and are
therefore less than ideal for growing large quantities of the
virus from which a live vaccine has to be prepared. However,
the experiment has shown that it is at least possible to do this
on cell culture, thus sparing the marmoset. Because of the many
human sufferers from this worldwide disease, it is hoped that
eventually a suitable cell will be found to make vaccine preparation a reality. (Provost,P.,1979).

"Recycling", or Repeated Use
of the Same Animal

There is something repulsive in the idea of keeping animals alive for as long as they can "last out" during a series of
surgical or other procedures. At least in the form of "practice
surgery" this has been banned in Great Britain since the Cruelty
to Animals Act came into being in 1876, although there *is* a loop-
hole: if what is being done can be described as an "experiment,"
it is legal. Unfortunately, recycling of primates has been vig-
orously advocated in the U.S. ever since the supply here began
to diminish (30,000 available in 1978 as against 55,000 in 1973).
Experimenters say they have to do it in order to conserve endangered species, and because primate models have become so expensive. Thus in a 1974 statement from the Division of Research

Resources, NIH, investigators were asked "to use their nonhuman
primate subjects to the maximum. Whenever feasible, scientists
should use an animal for several studies." (Anon.,1974,p.1).
An exhortation to do likewise, "as monkeys become more expensive," appeared in the NIH-sponsored ILAR *Nonhuman Primates*
(ILAR,1975,p.12). However, in this publication there is some
recognition of the need for improved caging to make the lives of
the primates being subjected to this high-pressure research more
bearable. It suggests

"...new designs for laboratory holding cages that will
accommodate the changing patterns of primate research, particularly the increased length of time for which animals are held and
the increased frequency with which both long-tailed and arboreal
species are utilized. (For example, sliding doors between com-
partments would provide convenient means for isolating individual
animals for examination, false floors of proper height above the
substrate could prevent the tails of long-tailed species from
becoming soiled with water or feces, and the addition of perches
would allow for normal foot and tail postures.)" (*Ibid*.,p.8).

The 1977 National Primate Plan continues to urge "multiple
use and recycling" and proposes a "users' service" which will
facilitate the transfer of primates to other facilities when the
original laboratory has run out of ideas. (IPSC,1977,p.13).

A painful dilemma - subjecting one group of animals to maximum stress in order to save a larger group from another kind of
stress, or even destruction. The only solution is a renewed
effort to help both groups by pressing the hunt for alternatives
to replace this "end justifies the means" type of exploitation.

Physical Restraint

One of the primate's misfortunes is the fact that he
shares with humans an unwillingness to be tethered like a cow
or leashed like a dog. This means that in many experiments he
must be under physical restraint, sometimes continuously for

months, in a device which reduces this lively and playful creature to straightjacketed near-immobility (Fig. 1). Even worse for
a social animal is what frequently follows: isolation in sound-
shielded cubicles - boxes 3 to 4 feet high, about 2 feet wide
and only deep enough to allow a primate chair to be rolled in.
(BRS/LVE,1976). Here are excerpts from two typical experiments:

1. Harvard Medical School (M.C. Moore-Ede and J. Herd):
"The studies were performed using six adult male squirrel monkeys....For periods of up to 3 wks. continuous urine collections
were obtained from unanesthetized monkeys, conditioned to sit in
a specially designed metabolism chair within an isolation chamber....The monkey sat on a bar and was restrained by a Plexi-
glas sheet which served as a table around its waist." The
experiment traced the daily rhythm of excretion of water,
sodium and potassium in chaired monkeys. (Moore-Ede,M.,1977,
p.F128-F129).

2. Walter Reed Army Institute of Research (J.W. Mason):
Fifteen monkeys were placed in restraining chairs, and each
chair was then isolated "inside a sound resistant, dimly illu-
minated, blower-ventilated booth. Urine collection was initiated immediately upon placement of all monkeys in the restraining
chair and was continued throughout the full 8-week period that
the monkeys remained in the chair." The experiment documented
the monkeys' emotional response to the stress of chair restraint
by noting the increase in cortiocosteroid excretion. This increase is associated with a physiological stressful situation.
(Mason,J.,1972,p.1292).

3. Yerkes Primate Research Center, Atlanta, Ga.: Studies
on the wasting of muscles caused by restraint and inactivity
in rhesus monkeys. (Anon.,1977a,p.1).

Although there are certain procedures, for instance an experiment such as No. 1 which involved a continuous metabolic
study, and No. 3, which by design require chair restraint without
a break, primates are sometimes housed in chairs for long periods
merely to avoid the bother of capture in the home cage and
transfer to and fro. However, techniques now exist which make

such permanent restraint unnecessary in most cases. Glassman,
Negrão and Doty at the University of Rochester describe a procedure by which macaques can be kept in their home cages yet
easily and safely put into primate restraining chairs when they
are needed for testing. The chairs are designed so that they
can be attached when required to the door of the cage where the
monkey is living, and the monkey is drawn into the chair by
means of a chain which, in a protective plastic tube, encircles
his neck. (Glassman,R.,1969).

Nevertheless, in studies of bone loss as a result of total
body immobilization, or of weightlessness in space experiments,
monkeys may be immobilized for a matter of months in restraint
systems. W.H. Howard and associates, who, in a National Aeronautics and Space Administration study on bone resorption, kept
monkeys immobilized for 10 weeks in a kind of nylon straightjack-
et on an aluminum frame, summarized some of the alternate methods.

"Nerve sectioning has been used by some workers; however,
because of its essentially irreversible nature, it was considered inappropriate for the present studies. Splints and
casts have been used, but they are inconvenient for routine examination for abrasions, sores and general body condition. With
a cast there would still be a problem of housing the animal on a
stand or carriage with attendant accessory equipment. Pillory-
neckplate type restraining chairs used in behavioral studies are
constructed of hard materials and would seem at the outset not
well suited for long-term restraint. Indeed, these have been
reported to cause edema, psoriasis, and decubital ulcers when
the animal was restricted to a sitting position." (Howard,1971,
p.112).

Doty, in an editorial in *Experimental Neurology*, also comments on this:

"In very few, if any, instances is it necessary for the restrained animal to be forced into the abnormal posture of sitting
in the style of Western man or of holding its chin at 90° with

respect to the spine; yet many primate restraining devices thoughtlessly impose this posture on their occupants." (Doty, R.,1975,p.ii).

Finally, I quote comments of Donald Barnes, a psychologist who worked for 16 years, up to 1980, at the School of Aerospace Medicine, Brooks Air Force Base, and became very critical of the experimental use of primates there. Speaking of restraint in behavior modification by electric shock, he said:

"The restraint devices used are barbaric in themselves: e.g. metal couches with metal neck, belly, and ankle restraints. As the animal struggles to free itself, it often loses its teeth to the neck bar, gains severe abrasions on the abdomen (often wearing entirely through the abdominal wall), or so severely chafes its ankles that they bleed and become infected; and the animal is shocked and shocked again (sometimes hundreds and hundreds of times per day), until it either does the ex-perimenter's bidding or is 'flunked out' to another program requiring no training." (Anon.,1980f).

Neurophysiological Research

Neurophysiological studies accounted for 16% of the experi-ments in the ILAR survey. These are experiments which investi-gate any of the special senses (seeing, hearing, etc.) or the brain centers controlling them, or the central nervous system and its affiliated nervous networks: peripheral, somatic and autonomic, or the neurochemicals like norepinephrine and ace-tylcholine which act as neurotransmitters. The primate, because he is cousin to man, with a brain and nervous system similar in many respects, is a favorite target of the neurophysiologist.

Recording Muscular Movement Registered in a Single Brain Cell

In June 1974 I visited the Laboratory of Neurophysiology at the National Institute of Mental Health, Bethesda, Maryland and talked to its chief, Dr. Edward V. Evarts. A thin, keen-appear-ing scientist with a Harvard background, he was investigating the brain mechanism controlling muscular movement in monkeys.

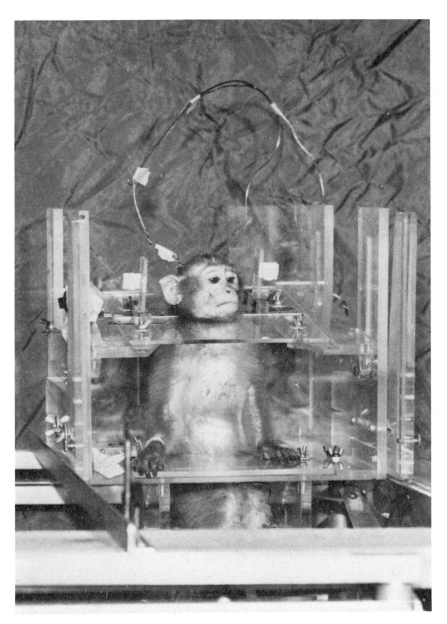

Fig. 1. Monkey in restraint chair. (Courtesy: International Primate Protection League).

His technique was to insert permanent stainless steel bolts
through the monkey's skull, one on the left and one on the
right side, and a steel cylinder through the top of the skull.
The latter allowed access to a microelectrode so small in cali-
ber that, while the monkey performed a series of wrist movements,
the instrument was able to record the activity of single nerve
cells in the motor cortex and other regions of the brain. During the manipulation, the monkey was restrained in a chair with
his head immobilized by the skull bolts attached to clamps on
either side.

The monkey was trained to make the desired wrist movements
through responses to a signal light; correct responses were re-
inforced by rewards of fruit juice delivered through a mouth
tube. Dr. Evarts said that in this procedure monkeys can be
trained by rewards but not by punishment. Some species cannot
be trained at all; he recalled one monkey of a particularly affectionate type which became very hurt when they attempted to
restrain him. The rhesus, while often aggressive, responds
well under the sympathetic management of a good animal handler.
Sometimes they would go by themselves to sit in their chairs
(anticipating, I assume, the juice reward). I saw one merely
restrained in a chair - "getting accustomed to it;" - when I
looked in he bared his teeth at the unfamiliar presence but when
I passed again on the way out he appeared to be asleep. As for
the steel bolts, Dr. Evarts maintained that these were no worse
than those implanted in the necks of people who had crushed cer-
vical vertebrae.[1]

[1]
*Also, Dr. G.H. Bourne, when he was Director of Yerkes Primate
Center, is quoted as saying that the permanently implanted electrodes are not irritating, since monkeys do not attempt to remove
them. "If you had something there that was causing pain, the animal would be scratching and pulling at it." (Stilley,F.,1975,
p.74).*

The monkeys used were young - preadolescent or adolescent. They never grew old, being killed after one year of life to permit examination of the brain tissue.

In an article in *Scientific American* published the year before my visit, Dr. Evarts described his experiments and suggested that they were producing a better understanding of the role of the deeper brain centers, such as the cerebellum and basal ganglia, and their relation to the cerebral motor cortex. Since these deeper centers are implicated in motor disturbances in humans, for instance in Parkinson's disease, he suggested that "a particularly promising area of investigation for the near future is the analysis of experimentally produced motor disturbances in monkeys." (Evarts,E.,1973).

This courteous scientist in his attractive office with Japanese prints and a Harvard chair, speaking of his appreciation of the personality differences of his monkeys and then introducing me to his kindly technician, tempted me to feel that the lot of these primates, compared to that of many other experimental animals, was not too bad. But when he writes about "experimentally produced motor disturbances," it reminds us that there is another aspect to neurophysiology - one which does not content itself with the mere recording through microelectric probing of relatively insensitive brain units.

Experimental Mutilations of the Nervous System

This other, darker side of neurophysiology is largely concerned with the effects of mutilations of various parts of the nervous system to see what functional disturbance results. Cats, because of their cheapness and availability, are more commonly used, but primates are by no means neglected. Since *any* mutilation (by deliberate surgical or electrolytic damage) causes some loss of function, there is always something to describe, and the variations are infinite.

At Yale University Medical School, the home of much brain mutilating "aggression" research (see p.20,41), Y.H. Huang caused electrolytic lesions in the locus coeruleus (a brainstem structure) of nine stumptail macaques. The locus is one of the "pleasure centers," and is so-called because gentle electrical stimulation of this area is remarkably titillating. It is rich in norepinephrine, one of the neurochemicals associated with stress and emotional reactions, and, sure enough, when all the monkeys were killed a month later, norepinephrine in the damaged area was much diminished. (Huang,Y.,1975). The experimenter had no hesitation, apparently, in sacrificing the lives of all these primates. Was there no alternative? M. Schlumpf and his associates may have found one: they have succeeded in growing norepinephrine-containing neurons of the locus coeruleus of the rat's brain in culture. The authors state that norepinephrine and dopamine levels can be measured by radio-enzyme assays *in vitro*, and claim that these cultures are well suited for neurophysiological recording, as well as morphological, biochemical and pharmacological experiments. (Schlumpf,M.,1977).

Alas for Yale's nine stumptail macaques!

Another fairly common procedure is directed not at the brain, but at the afferent nerves which transmit sensation from the limbs to the brain. These nerves are cut, "deafferented," in various limbs of young, newborn or fetal monkeys to see what effect it will have on their movements in later life. (Taub,E., 1976). The public can learn about the operative procedure and the crippled monkeys which result in a film produced (1968) by

the National Society for Medical Research, ANIMAL SECRETS - HOW
THE MIND BEGINS. In the film, inexplicably narrated by the
writer-philosopher Loren Eiseley, the deafferentation of a monkey fetus is hailed as an "historic" advance, likely to aid (in
some vague way) in the understanding and prevention of birth
defects and mental retardation. However, as frequently happens
in films vaunting medical research, and not infrequently in
scientific papers, one gains the impression that it is the ingenuity of the operative technique which is the *raison d'être*
for the presentation rather than any tangible therapeutic benefit.

On the other hand, the *blinding* of primates is not a procedure publicized in films promoted by defenders of animal experimentation. In this book, examples have been given in the
chapter entitled, "Behavior: Deprivation Experiments," p.47-48.

Behavior Modification

Finally we come to behavior studies - not the comparative
few involving observations in the field along ethological lines
as conducted by Goodall, Schaller and others, but behavior modification in the laboratory, which more accurately might be
called behavior distortion.

This subject, including the use of primates, has already
been discussed in the chapters on behavior, p.31-80, but
here are two further illustrations.

Donald Barnes, whose comment on physical restraint has
already been mentioned (p.128), left the School of Aerospace
Medicine at Brooks Air Force Base, San Antonio, in 1980,having
become increasingly critical of the inhumane experimental work
there. As a psychologist, he had participated for 16 years in
experiments subjecting primates to ionizing radiation injury and

and to nerve gas (organophosphate) poisoning - an investigation
which has recently become more interesting to the Air Force than
radiation studies. The onset and degree of poisoning is measured by impairment in the animal's performance of behavioral
tasks. After leaving, Mr. Barnes made a statement to the International Primate Protection League, of which the following is
an excerpt.

"I can no longer perform experiments with animals doomed,
by virtue of their participation in such experiments, to a very
early death, if not to pain and suffering, during the final weeks
and months of their existence.

From 1966 until mid 1978, I performed innumerable experiments with rhesus monkeys, and 2 or 3 such experiments with
baboons. In each experiment, 6 to 12 subjects were trained by
the use of electric shock to perform a task of human design,
i.e., not within the primate's normal behavioral repertoire.
It is no simple task to train a rhesus monkey to complex visual,
auditory, and tactile-kinesthetic discrimination. Although the
papers written to report such experiments claim that very low-
level shock is utilized as reinforcement, (3-5 ma), such statements are simply untrue. It may be that 3-5 ma is sufficient for
maintenance of acquired behavior, but such current levels are far
below those required to initiate early responses approximating
the desired behavior.

The shock generators are designed and manufactured by BRS
(Behavioral Research Systems) and deliver at least 50 ma at 1200
volts. I couldn't even guess at the number of times I've seen
these units used at full power to punish a slow learner or to
otherwise 'reinforce' undesirable behavior: well into the thousands; however, the learning process is replete with other dangers for the monkey as frustration leads to other self-destructive
behaviors, e.g. biting hunks of meat from an arm or hand, pulling
out hair until the subject is bald in accessible spots."

Once it has been trained in this fashion, the animal is
irradiated or fed the toxic substance, and the onset and degree
of poisoning is measured by impairment in the animal's performance of the behavioral tasks - often by its decreasing ability
to avoid repeated, painful electric shocks. (Anon.,1980f).

For other experiments of this type, using strong electric shocks, see p.138.139.

Communicating with the Apes

After recording this long saga of abuse of primates, it is a relief to recall that there are some very remarkable experimental alternatives which have opened up new insights into primate mentality and behavior. These are the experiments mentioned on p.110, teaching primates to communicate through various symbolic methods. Two fine films have been made on the work with the chimp Washoe; for reviews, discussion and distributor, see the Argus Archives publication, *Films for Humane Education*, (Scott,R.,1979). A film has also been made on Penny Patterson's work with the gorilla Koko. Koko has been taught to communicate through sign language and has a vocabulary of several hundred words. The sophistication of some of these exchanges is illustrated by the following anecdote:

"Three days after Koko had bitten Patterson in a fit of anger, she asked the ape (in sign language), 'What did you do to Penny?' Koko replied, 'Bite.' 'You admit it?' Patterson asked. Looking a little contrite, Koko said, 'Sorry bite scratch.' Patterson then asked Koko why she had bitten her. 'Because mad,' came the answer. 'Why mad?' 'Don't know.' The conversation ended." (Leakey,R.,1978,p.53).

A large literature has appeared on this research, which has implications not only for nonhuman primates, but also for a better understanding of the development of the brain and language in human primates. And yet, at the very moment when we seem to be on the verge of opening an astounding new channel of communication with our nearest relatives in the animal kingdom, the relentless population decline among the apes is accelerating. As Eugene Linden, in his excellent book, *Apes, Men and Language*, puts it:

"It would be unutterably sad to let any of these animals disappear. After so long, we have a lot to talk about. Perhaps they are the first and last creatures who can tell us who we really are." (Linden,E.,1975,p.196).

"Hitting, Pinching, Pricking, Slapping, Shocking" etc.

But this benign and civilized research commands only a minute proportion of the funds spilled out of the federal purse into the eager hands of the generality of primatologists. The kind of experiments which many of them prefer is illustrated by the following retrospective bibliography (1910-1974) of 582 stress and aversive (fearful or punishing) behavior experiments on primates. It was not compiled as "ammunition" for anti-vivisectionists but, on the contrary, was apparently intended to aid (and perhaps suggest ideas to) investigators engaged in similar research. The aversive events which these sensitive animals are subjected to include:

"air (pressurized and wind), alien species (human, human staring, and snake), aversive drugs, crowding, darkness, electrical stimulation of the brain, gravitational forces, hitting, isolation, looming, noise, nonreward (frustration), pinching, pricking, radiation, restraint, sandpaper, sensory deprivation, shock (electric), slapping, social defeat (induced experimentally), strobe light, tastes (foul), temperature extremes, threat of social separation, time-out from positive reinforcement, unavoidable aversive events, unpredictable stimulus changes, vertical chamber confinement, visual cliff, and water (rain)." (Stoffer,G.,1976,p.18).

Eight

RADIATION

"No Pain, No Distress, No Drugs" -
and No Description

In their annual report under the Animal Welfare Act for
1976, Battelle Pacific Northwest Laboratories listed the use of
2,538 animals of species reportable under the Act. Battelle
specified that none of these animals, with ten exceptions, suffered any pain or distress nor needed any pain-relieving drugs.
Since the Act requires no substantiation of the "no pain - no
distress - no drugs" listings and, very regrettably, no description of the procedures involved, one can only marvel at
how a large laboratory can use animals in the many ways one
knows they *are* used and keep them all free of any distress.
(USDA/APHIS,1976,Battelle Memorial Inst.).

Gastro-intestinal Poisoning
from Radionuclide

The ten unhappy few among these were a group of beagle
dogs noted under the heading "Pain-no drugs," and for this
listing the laboratory's attending veterinarian, Stephen E.
Rowe, supplied the required brief explanation. It is reproduced here to illustrate a test of toxicity of a radioactive
substance - a type of testing to which many animals have been
exposed and sacrificed over the years:

"In experiments to obtain dose-related data to predict
the gastro-intestinal effects and gut-related radiation

syndrome of accidental ingestion of a strong beta emitting
radionuclide in man, 10 beagle dogs were given oral doses of
106 ru which caused gastrointestinal lesions, diarrhea and
death. Analgesics were not used because they could be expected to alter gastrointestinal motility, passage time of
106 ru in the intestine, radiation dose to the gastrointestinal tract and other body functions to an extent that reliable dose-related data could not be obtained. To minimize
pain the animals were sacrificed when clinical signs indicated
death was imminent. To minimize the numbers of animals required
for the experiments, the dose groups were completed in sequence
so that data from one dose group could be used to plan sequential dose-groups." (*Ibid.*).

Fatal Radiation Pneumonitis

Another experiment, at the Lovelace Foundation, Albuquer-

que, New Mexico, subjected young beagles to inhalation of par-

ticles of Yttrium 90. Thirty-seven animals developed radiation

pneumonitis and died from 7 1/2 to 277 days after exposure.

Postmortem examination of the lungs of dogs who died after an

acute illness showed an inflammatory reaction with vascular con-

gestion, accumulation of fluid and occasionally tissue destruc-

tion in pulmonary blood vessels. With pathology like this, the

dogs must have experienced increasing shortness of breath,

coughing, choking and finally a suffocating death. (Slauson,D.,

1976).

Fifty-four Monkeys in the Electric Chair

Since everyone nowadays is nervous about radiation expo-

sure, there is plenty of money available for animal research in

this area, especially through the Defense Nuclear Agency. For

those investigators whose predilection is to rely heavily on

conditioning via punishing electric shock, this must be a boon

as well as a bonanza. It is not as easy as it used to be to

shake the money tree for support of interminable rat-shocking

experiments whose only justification is a vague promise that
they will help us to "understand aggression." But by devising
a procedure whereby an animal can be trained to escape painful shocks, then irradiated to the point of incapacitation...
yes, the public will buy this because of the radiation angle,
and even accept monkeys as the victims rather than the despised
rat.

A. Bruner and associates at the Lovelace Foundation are
prominent in this endeavor. In a typical experiment, they confined 54 rhesus monkeys to plastic restraining chairs, with
heads also in restraint, from Monday morning to the following
Friday afternoon. Electrodes were attached to the legs, and
different intensities of shock applied to the feet, either "weak"
shock, 6-13 milliamps, 0.5 sec., 60 Hz., "so as to produce a
clear emotional reaction but without vigorous escape movements,"
or "strong shock, 10-15 mA, 0.75 sec., 60 Hz., set to evoke
strong escape movements and vocalizations." (Parenthetically,
15 mA is excessively strong and painful shock - 4 to 5 mA is
painful to humans, more than 6.5 is "unbearable." (Weisenberg,
M.,1975,p.171). It seems incredible that such agony could be
inflicted on a small monkey in restraint merely in a "training"
procedure). (Bruner,A.,1975)

When the monkeys were sufficiently "shaped" in conditioning procedures to avoid foot shock by pressing a plastic but-
ton which interrupted the circuit, they were exposed to different levels of radiation to the whole body. In the next hour,
like the rats in the previous experiment, they gradually lost
their ability to escape from shock, progressing from "performance decrement" to "early transient incapacitation" in proportion to the dose of radiation received.

Blood pressure and heart rate were measured in the experiments by means of a femoral arterial catheter extended into the abdominal aorta. This had been surgically implanted some days before irradiation, no doubt adding to the physical distress of the immobilized animals and, in fact, in several instances confusing the results of the experiment, since "a few of the animals were already debilitated prior to irradiation as a result of difficulties with their surgical implants." To this variable must be added two more: the usual retching and vomiting which naturally distracted some monkeys from their shock-avoiding maneuvers, and fear, which although totally ignored in the report, must inevitably have been caused by the painful nature of the monkeys' task.

In spite of the variables, the author satisfies himself with the aid of a complex "x^2 contingency analysis" that lowered blood pressure shortly after irradiation "is a necessary but not sufficient condition for the occurrence of performance decrement [and/or] early transient incapacitation" in these monkeys. (Bruner,A,1977).

Experimenters Ignore Effects of
Pain and Fear

Aside from the objection that the above experiments may be adding very little to what is already known about performance decrement in humans who have been accidentally exposed to various forms of acute irradiation, there is a real objection on scientific as well as humane grounds to the linking of the monkeys' task to painful electric shock. That pain and fear are variables which contaminate experimental results has been mentioned elsewhere in this book. I have cited Harold Hillman, S. Iverson and A. Heim, among others, and their criticism of procedures which gratuitously introduce these stresses. At the

very least, procedures could have been devised in which the task
is shaped by the withholding of reward rather than the avoidance
of punishment, or through conditioning by positive reinforcement.
Or, the performance decrement might have been observed in disturbances of normal physiological or behavioral rhythms (cf.p.77).

"Do You Have Any Idea How Miserable It Is to Die from Radiation Injury?"

Incidentally, in his 1977 experiment Bruner mentioned that
one of the monkeys who received a high dose of radiation was
permanently incapacitated and in fact died within an hour after
exposure. The "difficulty with the surgical implants" may have
been partly responsible, or perhaps the effect of the shocks:
the investigator does not elaborate. However, a single death
pales beside reports, from the USAF School of Aerospace Medicine,
Brooks Air Force Base, San Antonio, Texas, of a long-term project in which 501 two-year old rhesus monkeys were exposed to
proton radiation. Over a nine year period following irradiation
this resulted in 279 deaths from chronic pneumonia, gangrenous
skin necrosis, gastritis, acute leukemia and other painful conditions. (Krupp,J.,1976). Even more monkeys perished - 1,379 in
the years 1972-1977, according to the *Washington Post* (June 22,
1977) - in tests simulating the effects of the neutron bomb at
the Armed Forces Radiobiology Research Institute in Bethesda,
Maryland.

Donald Barnes, the psychologist already cited on p.128
and 133-134, who participated in experiments like the above at
Brooks Air Force Base from 1964 until he left in 1980, commented
as follows to the International Primate Protection League:

"Assuming the animals survive training (and many of them do
not), my job was to determine their resistance to ionizing radiation, i.e. neutron, gamma, flash X-ray. In years past, I was
ordered to keep a death watch on these irradiated subjects, which

meant, simply, to see what happened until they died of radiation injury. Do you have any idea how miserable it is to die from radiation injury? I do, I've seen so many monkeys go through it.

At any rate, I finally got permission to sacrifice my subjects after the experiment proper was completed (from 1 to 12 hours as a rule). We injected them with a compound designed to slow the heart gradually, thereby supposedly minimizing pain. I often did this myself in order to minimize suffering occasioned by clumsiness or ineptitude of technicians: on each occasion, I felt more strongly that I didn't have the right to kill these innocent creatures. As I became familiar with the use of the data gained from these experiments, I discovered that the data was not used to help Man in the struggle against his environment...the data was (and is) used to generate more worthless experiments, thereby killing and crippling more animals. I finally objected to doing any more experiments in this area." (Anon., 1980f).

Monitoring the Human Mutation Rate

Some very pertinent comments on this subject were made by Norman Anderson and his son Leigh, respectively director and staff member of the Argonne National Laboratory's molecular anatomy program, in a letter to *The New York Times*, written April 20, 1979:

"Potential increases in the rate of mutation and the incidence of cancer and birth defects are the central concerns arising from nuclear energy, medical X-rays and environmental pollutants. A basic common denominator of these effects is genetic injury. We do not know if the human mutation rate is being increased, because it is not being measured."

Instead, they point out, the policy has been merely to measure the radiation received when humans are known to have been exposed, and to deduce possible effects by testing animal models similarly exposed.

"Unfortunately, men are not mice, and mouse data is now found to be of little value in pending court cases. No animal is exposed to the same schedule of all the pollutants (dietary and environmental), self-inflicted insults, drugs, X-rays or the same microclimate to which we are exposed. The sum total of these is

what is of interest, and it decides the fate of the individual
and of future generations....New advances now make possible human
mutation rate measurements."

Possible changes in it, they add, will provide an index of
potential increases in both cancer and birth defects. (Anderson,
N.G.,1979). To achieve this form of epidemiological study, large
populations of two million or more have to be monitored, since the
incidence of gene mutations in normal populations is very low,
and these mutations are subject to periodic fluctuations in fre-
quencies - thus data have to be collected over a number of years.

The changes looked for are variants of proteins indicating
variant genes; chromosomal abnormalities in the blood of the newborn, and congenital anomalies such as achondroplasia (dwarfism
with short legs and normal trunk). (Grice,H.,1975). Once the
background mutation rate has been determined, increases in it
can be detected and the causes of these increases searched for.

The Checkered Human Condition Cannot be Copied in the Laboratory

In Apr. 1979 an article appeared in *New Scientist* entitled
"How Dangerous is Low-level Radiation?" It was written by K.Z.
Morgan, Neely Professor in the School of Nuclear Engineering at
the Georgia Institute of Technology, Atlanta. (Morgan,K.,1979).
Its message was that the risk of developing cancer from low-level
ionizing radiation was much greater than had been imagined, and
that the medical profession was responsible for over 90% of the
man-made radiation delivered to the public. Professor Morgan's
data have been largely derived from epidemiological and clinical
studies on humans, including observations on the survivors of
Hiroshima and Nagasaki and cancer statistics on workers in nuclear plants. He has reservations about the thousands of experiments conducted on animals and extrapolated "perhaps brazenly or
at best with misgivings" to man. He mentions various reasons

for these misgivings. Not only does the response to a given dose
of radiation vary widely from species to species, it can also
differ markedly within a species: for example, sensitivity to
leukemia induction, and life expectation once the disease is established, vary greatly with different kinds of mice. Furthermore, these observations derive from carefully controlled, inbred
healthy animals, whereas "man is a wild or heterogeneous animal
living in many types of environment with various eating and drug
habits, with many diseases and eccentricities, of various ages,
and so on."

This checkered human condition cannot be copied in the laboratory, yet it seems increasingly probable that the onset of a
malignancy or indeed other disease may be the consequence of many
more events than merely a measured dose of radiation such as is
delivered to the thousands of beagles, monkeys and mice under experiment. For example, says Morgan, "a given type of leukemia
may require as many as three successive events (like throwing
three electrical switches connected in series). Some of these
switches may be thrown by viruses, bacteria, chemicals, mechanical damage or radiation."

Prevention as an Alternative

In his recommendations for action Morgan says nothing about
additional testing of animals since he considers much of it inapplicable. Instead, he emphasizes research programs to define
more accurately the risk from human exposure to ionizing radiation.

Although Morgan does not describe these it should be noted
parenthetically that, as an alternative to the experiments which
seek to demonstrate the pathology occurring in the fetus or in
germ cells through irradiation of the whole animal, it is now
possible to study the potential damage of irradiation in humans

through the use of cell cultures. The irradiated cells can be
examined through an electron microscope and minute chromosomal
aberrations detected. The nature of developmental defects can
thus be predicted. (Federoff,S.,1975).

    But Professor Morgan's main thrust is toward prevention, in-
cluding the reduction of exposure from all sources, particularly
the nuclear energy industry, and above all from the medical use
of radiation. He estimates that " a reduction of only 1% in
unnecessary diagnostic exposures in the U.S. would reduce the
population dose of man-made radiation more than the elimination
of the nuclear power industry to the year 2000."

Nine

SURGICAL EXPERIMENTS

1. BURNS

Number and Type of Experiments

Everyone has experienced the pain of thermal burns, so naturally the thought of experimental burns affecting animals is very disturbing. The first question that comes to mind is: how numerous are these experiments?

Since government regulations do not call for a breakdown of the nature and purpose of experiments reported under the Animal Welfare Act, and since the government-supported National Research Council has failed to supply this information in its broader surveys of animal research, it is impossible to quote statistics of burn experiments for the U.S. However, Great Britain does produce some figures on this subject in the reports presented annually to Parliament by the Home Office. (UK Home Office,1978). A summary may be of interest.

The figures cited in *Table 1* are for the year 1977. An animal is counted as one experiment, and an experiment involving a number of animals is counted as many times as there are animals experimented upon.

If burning and scalding experiments are one-twelfth of one percent of the British total, the same percentage of the estimated 100 million American experiments (Fox,M.,1979,p.15)

Nature of Experiments: Burning or
Scalding by Any Means

Use of anesthesia	No anesthesia	For part of experiment	For whole experiment	Total
No. of Experiments (i.e. Animals)	2,120*	3,659	782	6,561

Purpose of Experiment	Study of normal or abnormal body structure or function.	Study or development of medical, dental or veterinary products.	For other purposes.	Total
No. of Experiments (i.e. Animals)	1,285	5,223	53	6,561

Species of Animal	Mouse	Rat	Guinea Pig	Rabbit	Other Vertebrate	Total
No. of Experiments (i.e. Animals)	1,917	2,249	2,329	3	63	6,561

Number of Experiments (All Types)

Total	Burning: Percent of Total
5,385,575	0.0012 (one twelfth of one percent)

Table 1. Burning or Scalding Experiments on Living Animals in the United Kingdom in 1977. (UK Home Office,1978).

*Exposed to ultraviolet rays simulating sunlight.

would total about 120,000. The British statistics reveal that
some 80% of the burn experiments involve the study and development of pharmaceutical products - presumably drugs designed for
use in the therapy of burns, although sunlamps may also be included in this group. The 2,120 experiments which were conducted, without anesthesia, testing ultraviolet light burn, must
have been part of this 80% and must chiefly have had to do with
the evaluation of anti-sunburn products and with the many drugs
which, even when taken internally, are photosensitizers; i.e.,
cause skin reactions in sensitized persons exposed to sunlight.

Although some of this pharmaceutical research may get into
the pharmacological journals, much of it is probably in-house
research and not reported in the literature. However, an occa-
sional reference in compliance with the Animal Welfare Act may
appear in an annual report of one of the drug companies; e.g.,
Johnson and Johnson's description of 52 burn experiments in rats
with the comment: "No postoperative analgesics were used because
it would interfere with experimental results." (USDA/APHIS,1973,
Johnson & Johnson Res. Foundation).

Postoperative Analgesia Is
Often Inadequate

To inflict a burn on an animal inevitably produces pain and
usually much postoperative suffering. One would expect that
those undertaking such distressing procedures would try to relieve their subjects through the use of carefully chosen anesthet
ics and analgesics. It is true that all the acute burning is
done under anesthesia. In a group of three experiments involving
guinea pigs, all conducted by Robert Wolfe and associates at
Louisiana State University Medical Center and at Harvard Medical
School, the following anesthetics were used:

1. "Whiffs of halothane before being burned and for approximately 5 min. afterward which was long enough to ensure that they

felt no pain (as indicated by the absence of squealing) before
lapsing into shock."

    2. "Whiffs of halothane."

    3. Halothane (2% halothane in 100% oxygen).

But none of the above (nor other burn experiments I have reviewed) refer to *effective* postoperative analgesia or tranquilizers. Apparently, these are not supplied, and no doubt if the experimenters were asked why, the reply would be that the drugs
might affect some of the physiological functions under investigation.

What about the whiffs of anesthetic gas continued for five
minutes post-burn in the guinea pig experiment No. 1 above? The
burns were produced by a three second immersion in boiling water.
Ten animals were burned over 70% of their body surface and seven
of them died within 24 hours. Another seven were burned over
55% of their body surface and all survived over 24 hours. The
purpose was to study cardiovascular and metabolic responses
during burn shock, but without employing anesthesia deep enough
to alter the response to the shock resulting from the intense
burn. Accordingly the animals were given "whiffs" of halothane
sufficient to prevent squeals of pain, but lasting only five
minutes and therefore not long enough to affect the physiological
functions being monitored. This cannot be considered effective
postoperative analgesia: after the initial shock had passed -
with some animals surviving into the second day - it seems certain that pretty severe suffering must have been the lot of many
of them, if not all. (Wolfe,R.,1976).

In another experiment a few months later, Wolfe and his
associates partially immersed 26 guinea pigs, "temporarily anesthetized with whiffs of halothane," in boiling water for three
seconds. Nothing is mentioned here about prolonging the anesthesia to avoid squealing; however, the animals all died between

8 and 24 hours post-burn. Various physiological functions were studied, and the results indicated "an important role of lactate in burn shock metabolism." (Wolfe,R.,1977a). In a third experiment, Wolfe and J.F. Burke produced a third degree burn by a much longer immersion (20 secs.) of 16 guinea pigs in boiling water, using 2% halothane in 100% oxygen. Five animals died between 60 and 72 hours post-burn, eleven survived more than 72 hours. (Wolfe,R.,1977b).

Further Examples of Burn Research

Other investigators, experimenting on rats using other than halothane anesthesia, also reported burns produced by partial immersion in boiling water or steam. No postoperative analgesia was mentioned. The purposes and findings were briefly as follows:

W.L. Brown and associates studied protein metabolism in rats who had been burned, and in others who had been burned and whose wounds were then seeded with the organism *P. aeruginosa*, source of an infection which is a major cause of death in severely burned humans. They discovered that the post-burn drop in concentration of the body's serum albumin is due to the large pool of this protein which forms in the wound area. (Brown,W., 1976).

J. Turinsky and colleagues found that a 20% surface burn injury in the rat is followed acutely by a brief period of glucose intolerance, succeeded by a much longer phase in which there is a secretion of large quantities of insulin, indicating that the hormone is striving to overcome some physiological resistance. The source of the resistance has not been determined. (Turinsky, J.,1977).

George Noble and associates burned 100 rats, followed by various procedures to determine the effect of heparin (which prevents blood clotting) on the vasculature of the wound and surrounding tissues. Heparin, given before the burn, was found to retard the devitalization of the blood vessels; given afterward, it had no effect. (Noble,H.,1977). This was a joint project of the University of Chicago Pritzker School of Medicine and the Yale University School of Medicine.

Repeated Burn Experiments on Sheep

Although guinea pigs, rats and mice seem to be the preferred species for burn experiments, R.H. Demling and associates at the Burn Unit of the University of Wisconsin Center for Health Sciences have experimented on sheep. Twenty-eight sheep were anesthetized with pentobarbital (10 to 15 mg/kg intravenously). A hind limb was scalded in 85° to 95°C. water, causing second to third degree burns. The second hind limb of each animal was burned in the same manner a week later, and heparin was injected at various times and in different doses to study its effect on edema (lymph leakage) in the flesh of the wound area. It was found that heparin did not have a beneficial effect on the edema. (Demling,R.,1979).

The fact that a second painful procedure was performed on each of these sheep just when it was recovering from the first may raise a question in the reader's mind. Is this humane? Known as "multiple-survival surgery," this practice, according to the National Institutes of Health "Guide for the Care and Use of Laboratory Animals" (which sets standards for research), is generally to be discouraged. However, the Guide goes on to say that such procedures may be permissible under certain conditions, including aftercare to alleviate postsurgical pain and when the operations "are related to components of a research or instructional project." (ILAR,1978,p.14). Since the experimenters make no mention of postoperative analgesics, one cannot be sure that the first condition has been met, but the fact that the repeated burning is part of the same experiment fulfills the second requirement, although it is doubtful whether this makes the sheep feel any better. Probably the Guide seeks primarily to discourage veterinary students or surgery interns from practicing a series of haphazard procedures on the same animal, accompanied by the

careless aftercare which is the usual hallmark of group callous-
ness and irresponsibility.

Synthetic Substitute: "Thermo-Man"

In order to study the way garments of different materials
ignite and transfer heat under varying conditions, the Du Pont
Company has been experimenting with "Thermo-Man," a life-sized
dummy made of epoxy-fiberglass resin, implanted with 122 sensors.
These detect temperature at a given time; and through a skin-
simulating computer program, the actual heat transferred from a
burning garment and the depth of thermal injury (first, second
or third degree burn) can be calculated for human skin. The computer program is sophisticated enough to account for the liquid-
to-vapor phase change that occurs at $100^{\circ}C$. in skin-tissue water,
and a controlled wind system can simulate Thermo-Man in motion.

Excellent correlation has been demonstrated between Thermo-
Man burn damage information and that obtained from injuries in
humans and in experiments on pigs. (Bercaw,J.,1977).

Non-survival Experiments on Animals

Critics of the use of animals for experimental purposes have
been known to describe procedures which sound horrifying, such
as dipping dogs or monkeys in boiling water, without always specifying whether the animals actually suffered pain. But if the
animals have been anesthetized throughout the experiment, and are
never allowed to recover consciousness, the procedure itself cannot properly be called inhumane. With Thermo-Man only the effect
of very intense heat - as from a flash-burn - can be demonstrated
there would be acute destruction of skin tissue *in vivo*. But whe
anesthesia can be prolonged for hours, in non-survival experiments, secondary changes can be studied in the skin and even in
other organs.

As an experiment with serious intent to contribute to burns research, and one in which animals did not suffer, I cite an investigation at the University of Gothenburg, Sweden, to study the effects of thermal injury on circulation and respiration in dogs. Seventeen dogs were used, and the burn was produced by immersing the rear half of the body of the ketamine-anesthetized animals under water at 80°C. The anesthesia was maintained by hourly infusions until the dogs were killed six hours later. The experiment yielded information about the effect of burn-produced denatured proteins on other components of the dogs' blood, and especially on one known as "complement." Although this substance helps in scavenging damaged and coagulated cells, it can also overact and cause hurtful side effects on the lungs and cardiovascular system. (Heideman,M.,1979).

Human Cell Cultures

Although skin cultures (epidermal cells) have been developed (Nardone,R.,1977), and cadaver skin is also available, burns research is largely concerned with the physiology of the whole organism, not merely with the traumatized skin. The use of *in vitro* techniques such as cell and organ cultures in this field is therefore rare. However, it will be recalled that in the British statistics (cf.p.146) almost a third of the experimental burns reported involved exposure to ultraviolet rays, although admittedly these were not considered painful enough to require anesthetics. An experiment at the Laboratory of Radiobiology, Harvard University, may be of interest in this regard. Williams and Little pretreated a culture of human cells exposed to lethal effects of ultraviolet light with the drug proflavine. These cells showed a significant enhancement in survival compared to untreated cells. The authors suggest that their cell culture technique could lend itself to an investigation of the various "photoproducts" induced in cells by ultraviolet light, and the role these photoproducts play in the effects observed in irradiated cells including cell death, mutation and transformation. (Williams,J.,1977).

Theoretically, it seems possible that the effects of toxic products of thermal burns from human patients might also be investigated in human tissue or organ cultures. If these cultures were kept alive and differentiated in some of the sophisticated media now available, the period of observation could be prolonged into days and weeks, thus reaching the third stage in the time scale of acute, subacute and chronic alternatives.

Postoperative Pain-relief

Since burns research concentrates ultimately on the physiology of the whole organism, such matters as the balance and distribution of fluids and electrolytes, the status of the vascular system and the blood, the degree of pain and shock, and the systemic reaction to drugs are all of vital importance. Although much has been learned from study of the human patient, there are still unanswered questions which impel investigators, with the approbation of most of the public, to turn to the animal model, with all organs intact and interacting, and without the obvious limitation of in vitro cultures. And a longer period of study is often required than is possible with the animal under general anesthesia in a non-survival experiment.

Recognizing, therefore, that such research will inevitably continue, we must concentrate more on the animal's pain and ask, can this be relieved if not eliminated?

First of all, the experimenter should give more recognition to this pain. When a human is burned, much attention is given to pain: to the reactions it may stimulate in the patient through the nervous and endocrine functions under its control, and to

its relief by drugs. To ignore similar reactions in suffering
animals, or to sensibly increase the suffering by withholding
pain relief, may add to the difficulty of comparing the effect
of the trauma in animals with that in humans (or compound the
difficulty which already exists owing to species difference).
The many physiological reactions of an animal in fear and pain
enumerated by Harold Hillman (cf.p.14), apparently ignored by
many investigators, are, according to Hillman, "almost certainly
the main reason for the wide variation reported among animals
upon whom painful experiments have been done." Possibly this
may explain the confused results reported in the Noble and
Demling experiments (cf.p.150) on the effect of heparin, which
are in conflict with earlier studies of the action of this drug.
These unexplained variations in findings always result in repeated experiments and the sacrifice of more animals. Demling
concludes his paper with the words, "Further studies are necessary...."

     Secondly, the experimenter should give pain relief *through-
out* the experiment comparable to that offered a human patient.
Comes the objection: pain-relieving drugs must be minimized or
withheld because they alter the physiological functions of the
animal and may "defeat the purpose of the experiment" (a stock
phrase parroted from numerous official research guidelines).

     This glib response spells indolence, ignorance, and in-
humanity. There is a perfectly good way to assess the effects
of any pain-relieving drugs which might be used: namely, by
running a control series of animals which would not be subjected
to the burn but would receive the anesthetic and/or analgesic.
In this way, any variations in the physiological status of the
animal caused by these drugs could be noted, and could be taken
into account when other parameters were being evaluated. As it

is, control animals are always included in these investigations, and this suggestion would simply add more of the same.

This should lead to more and probably much needed knowledge about the effect of anesthetics and analgesics generally and clarify some hitherto inexplicable variables. Interestingly, one of the anesthetics used in several of the above-mentioned experiments, halothane, has been studied in rat cell cultures, yielding information about its influence on protein and lipid synthesis. (Ishii,D.,1971).

A complementary approach to the study of these drugs is through mathematical modeling and the use of digital computers. Thus in an article entitled "Simulation as an aid to the replacement of experimentation on animals and humans," R. and M. Harrison discuss the value of computers in predicting the effects of drugs and combinations of drugs in various dosages. (Harrison,R.,1978). Unfortunately, many experimenters are ignorant of these techniques, but that is a poor excuse for a lack either of humaneness or scientific precision. Essentially, the investigator in these burn experiments needs to know the effects on body chemistry of various drug inputs. Here the computer can help by simulating the behavior of human blood at steady-state equilibrium, then reflecting it under various simulated stresses. And its ability to simulate the respiratory system could be valuable in conjunction with studies of anesthetics made on the living animal. (Maloney,J.,1963).

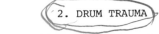

2. DRUM TRAUMA

*Battering Animals in a Drum: How
it Started*

A method of inducing experimental traumatic shock by battering the entire, unanesthetized animal was introduced by

R.L. Noble and J.B. Collip in 1942. T

volving drum, 15 inches in diameter.

tions, two inches high, are attached w

carries a rat or guinea pig up the sid

it to fall when it reaches the top and

following projection. To prevent an attempt by the animal to
break its fall, the paws are taped together. The drum is re-
volved by a motor 40 times, more or less, a minute. An animal
is subjected to two falls for each revolution of the drum, a
standard initial 'drumming' consisting of 360 revolutions, or
720 falls in nine minutes. The following are typical injuries
from the battering in the drum: teeth broken or knocked out;
bruising of head, paws and scrotum; hemorrhages into muscles;
bruising of liver; engorgement of bowels, kidneys, lungs and
intestines, with later appearance of ulcers in stomach and
intestines. (Bayly,M.,1952).

Since 1942, this method of traumatizing conscious animals,
chiefly rats, has been steadily used in the U.S. One investigator, B.W. Zweifach, began experiments with the drum at New York
University in 1943, and has carried them on into the 1970's. He
has had the help of several hundred thousand dollars of National
Institutes of Health grants - out of the taxpayer's pocket, of
course.

Aside from humane groups which have continually protested
against experiments which result in such extreme and prolonged
suffering in unanesthetized subjects, there have also been objections from scientists themselves. As long ago as 1949, six
distinguished British scientists wrote to the conservative medical journal, *The Lancet*, asking "What amount of suffering is
legitimate to inflict in the course of experiments on animals?"
and ended with an appeal to other scientists to condemn such

s the Noble-Collip drum as "shocking to a normal human ence." In a definitive survey of experimental shock h was extensively published in *Federation Proceedings* and ncluded a paper by Dr. Zweifach himself, H.B. Stoner contributed a major review, "Traumatic Shock Models." He flatly stated that the Noble-Collip drum had outlived its usefulness. "It is impossible to describe the extent of the injury and study the injured tissue quantitatively....The method seems altogether too crude for modern purposes." (Stoner,1961).

Still Battering in the 1970's

What was too crude by 1961 was nevertheless quite acceptable to New York University School of Medicine in 1970, where Z. Hruza and B.W. Zweifach, supported by the ever-helpful U.S. Public Health Service (National Heart and Lung Institute), reported further work with the Noble-Collip drum. One table shows that 75 rats were battered until they died, some having received adrenalin-type drugs and some none, but all, unanesthetized, having been subjected to anywhere from 3,000 to nearly 5,000 falls. Since it has been found that some animals, though not, of course, escaping injury, can adapt better than others to the trauma and survive for varying periods, they can be used to test the effect of various drugs calculated either to increase or decrease their resistance to shock. Thus they are repeatedly returned to the drum for more "tumbling," and are then sometimes operated upon (now at last anesthetized) for examination of their organs. (Hruza,Z.,1970).

Anesthesia - To Use or Not to Use?

Noble and Collip originally introduced their drum with the idea that it would produce a pure kind of shock, uncomplicated by anesthesia (their animals received no pain-relieving drugs), hemorrhage (as used in another type of experiment to produce

shock) or infection (as in endotoxin-stimulated shock). The
omission of anesthetics has been continued by some experimenters
up to recent years; thus in 1976 V. Lockard and R. Kennedy at
the University of Mississippi Medical Center were still battering unanesthetized rabbits 400 times at 30 r.p.m. in a Noble-
Collip drum. (Lockard,V.,1976).

On the other hand, perhaps because of some sensitivity to
the growing public outrage over such experiments, "light anesthesia" has made its appearance. At Albany Medical College,
New York, two groups, using rats in the drum, employed respectively "light ether anesthesia" (Sarfeh,I.,1977) and intraperi-
toneal sodium pentobarbital in a dose of 2 mg/100 g. body weight
- also "light," since the normal dose is 3.5 mg. (Kaplan,J.,
1976). At Downstate Medical Center, Brooklyn, New York, B.M.
Altura, battering rats at up to 800 revolutions in the drum,
used pentobarbital, 3 mg/100 g. body weight; this is as "light"
as the above because it was given intramuscularly, for which
the normal anesthetic dose would be higher than 3.5 mg. (Altura,B.,1976).

Briefly, the post-trauma effects under investigation in
the above experiments were as follows: reduced antibacterial
function of the phagocytes ("scavenger cells"): Lockard, Altura, Kaplan; functional abnormalities of the liver: Sarfeh;
greater resistance to lethal trauma of female rats compared
to males: Altura.

The fact that some anesthesia is being given (although
probably not enough to eliminate all pain) shows that the investigators now discount such drugs as contaminants of their
battering. If they are *not* contaminants then thousands of animals who have been denied pain-relieving drugs during the forty
years that this crude Noble-Collip instrument has been in use

have suffered to no purpose. What kind of "science" is this?
Furthermore, the animals still suffer severely after the effects
of the anesthesia wear off because, in no case, are postoperative analgesics administered. However, as was pointed out in
the discussion of post-burn analgesia on p. 155, the effect of
all these pain-relieving drugs could have been assessed long
ago by running a control series of animals which had received
the anesthetic and analgesic, or tranquilizer, but had not been
battered in the drum. Once these effects, if any, were known,
the assessment of whatever other reactions were under scrutiny
could have proceeded unhindered, and the animals could have been
protected from pain until they were euthanized.

Alternatives

As long as the accident wards of hospitals are full of
automobile crash victims, surgeons have a wealth of material
to study. It is inconceivable that they need any scraps of information which may be picked up from the hit-or-miss battering
of rodents in the Noble-Collip drum.

3. IRRITANTS OF THE DIGESTIVE TRACT

Mucosal Irritation and
Corrosion

A child who innocently swallows a highly corrosive substance like lye, or a strong acid, or an adult who does the
same by accident or with suicidal intent, create a catastrophe
for themselves and a very difficult problem for the surgeons who
must treat them. The mouth, the throat, the esophagus, the
stomach and the upper intestines are in the direct line of
march of these irritants, which can produce ulcers, perfora-
tions, disabling strictures and sometimes death. Strong alka-
lis are most damaging to the esophagus and strong acids to the

stomach and duodenum.

Against the background of these grim facts, household products containing lye or acids, and similar substances as defined by the Federal Hazardous Substances Act, are tested by their manufacturers on animals. This is done so that their toxicity can be defined preparatory to their being properly labeled, packaged and sold to the public. Also, the manufacturers are careful to protect themselves by extensive testing against potential lawsuits charging negligence brought by those injured by or claiming to be injured by the products. The government body which administers this act is the U.S. Consumer Product Safety Commission, and the Commission itself may test these substances.

The animals into whose mouths, esophaguses and stomachs the toxic substances are poured include rats, rabbits, cats, dogs and swine among others. The most devastating of these procedures is the test for corrosion of the esophagus, and in an authoritative publication of the National Academy of Sciences, *Principles and Procedures for Evaluating the Toxicity of Household Substances*, the following suggestion is made:

"The need for a special test for esophageal corrosivity has been questioned on the grounds that the customary battery of acute tests for oral toxicity, skin irritation, and eye irritation, when combined with information on chemical and physical properties, can provide reasonable presumptive evidence of a severe irritant or corrosive hazard on ingestion in the absence of empirical data. Indeed, the need for such an animal test might also be questioned on humane grounds." (NRC,1977,p.55).

This is a step in the right direction inasmuch as "humane grounds" are invoked, as a reason for questioning the use of esophageal testing, by a writer for the research "establishment," which rarely concerns itself with the subject of humaneness. But the battery of acute tests mentioned as substitutes are far from humane.

A more acceptable alternative sparing the intact animal
is needed.

Alternatives in Acute Testing
of Irritants

The mucous-secreting epithelial cells of the mouth are of
some value here: they can be easily removed by gently scraping
the lining of the cheek, and in the laboratory their response
to irritant or corrosive substances can be observed through the
microscope. (Smyth,D.,1978,p.135).

If the effects of deeper penetration into the wall of the
digestive tract need to be observed, it is possible to remove,
safely, through a biopsy tube, full-thickness mucosal samples
from the wall of the stomach or intestines. This can be done in
humans as well as animals. The specimens can be cultured in a
nutrient medium (Trowells T-8), with added antibiotics, for from
12 to 48 hours.[1] Besides surviving longer than the isolated epithelial cells mentioned above (which live only a few hours),
these small organ cultures contain many cells which preserve
their usual anatomic relations and carry on metabolic activities
including the synthesis of mucus and glycoprotein. (Trier,J.,
1976).

It is true that these organ cultures have been developed
to study normal gastrointestinal mucosal functions and intestinal diseases such as celiac sprue and ulcerative colitis, but,
in principle, there is no reason why they cannot also be used
in toxicological studies and in the testing of substances for
irritancy and corrosiveness. Of course the taking of biopsies

---

[1]
*Or much longer according to B.F. Trump et al. at the Carcino-
genesis Program, Given Institute, Aspen, Colorado, 1977. They
reported that human esophagus, colon and other explants can now
"all be cultured for weeks to months with maintenance of normal-
appearing epithelium." (Harriss,C.,1978).*

requires some surgical skill and the organ culture procedure has
to be technically precise. The *in vitro* approach also has certain obvious limitations: motility studies timing the movement
of the poison through the alimentary canal cannot be performed,
nor chronic studies. On the other hand, such processes as the
synthesis of protein can be followed by adding a radioactive
protein precursor such as (^{3}H)leucine to the medium and observing either its incorporation into macromolecular tissue protein
or the failure of synthesis owing to the toxic insult. Finally,
many specimens can be removed during a non-recovery operation on
an anesthetized animal.

Chronic Effects of Irritation and Corrosion

I have been discussing acute tests for irritants and possible alternatives. However, a far graver problem arises in seeking alternatives to experiments on the *chronic* effects of irritation or corrosion. In a typical procedure, animals are forced
to swallow known corrosives. Surgeons then study the injuries
that develop and test therapeutic measures either through drugs
or surgery. Observation of the unfortunate animals may last for
months. Experiments on animals related to conditions which
arouse maximum human concern – in this case, severe internal injuries following the swallowing of caustic substances by children and others - are often passed over in silence by those who
under less affecting circumstances would be very vocal about the
plight of the animals. However, since this book is focused primarily on the pain of animals under experiment, procedures of
the kind about to be described cannot be overlooked, no matter
how great their value to humans, if only becuse of the atrocious
suffering which animals subjected to them undergo.

A group of surgeons at the University of Arizona College

of Medicine opened the esophagus in 139 "mongrel dogs" under
anesthesia, instilled lye into a 5 cm. segment of it which was
tied off during the 60 seconds the caustic remained within, then
flushed it out with water. This produced a burn extending
through the full thickness of the esophagus in 77 dogs who survived for at least two weeks. Twelve dogs died with perforation
of the esophagus in less than two weeks; the rest had less than
full-thickness burns and were not included in the experiment
since their injuries were judged not likely to develop into
stricture of the organ, the object of the study.

The dogs with full-thickness burns were divided into groups
of 24, 26 and 27 dogs respectively. The first group received
no treatment; the second received injections of a steroid with
the property of reducing inflammation and scar formation. These
latter had their gullets dilated twice a week by a mercury-
filled "bougie." The third group were not dilated but were
treated with a drug known as a lathyrogen which has the effect
of producing a less rigid, more plastic kind of scar tissue.

Seventeen of the 24 dogs who received no treatment developed severe anatomical stricture of the esophagus, leaving an
opening with a minimum diameter of half a centimeter or less.
Only four of the second group (steroid plus bougienage) and two
of the third (lathyrogen) developed stricture, but these two
groups tended to have persistent - and presumably painful -
ulceration, apparently related to the mechanical abrasion from
repeated dilatation and, somewhat inexplicably, to the effect
of the lathyrogen. The first group, who were untreated, healed
more quickly and escaped the persistent ulceration.

The animals were allowed to survive a total of 12 weeks
from the operation until death, "unless they lost 15% of body
weight and were obviously starving" (presumably because food

could not pass the stricture).

The surgeons learned that the stricture caused by lye resulted from contraction of the full-thickness wound, not from the proliferation of bulky scar tissue, as had been previously thought. Also, they found that the effectiveness of the lathyrogen in preventing esophageal stricture is as great as that of "conventional" therapy (steroid plus bougienage), "with the distinct advantage that the hazards and difficulties of bougienage" - described as "unpleasant and often unsatisfactory" - "are obviated." (Butler,C.,1977).

Postoperative Analgesia Withheld

Although the lye was instilled into the esophagus under anesthesia, there was no comment about postoperative pain relief; only a reference to an earlier paper. Nothing, however, was mentioned there, but again there was a reference to a still earlier paper for a description of the procedure. This "buck-passing" illustrates a recurrent problem facing anyone trying to discover whether analgesics routinely available to humans afflicted with extremely painful conditions were given, as they should have been, to animal subjects. Unhappily, the final answer, in a 1973 paper, proved to be "no": after the burn the dogs "were fed a soft-solid diet for four weeks and had no other treatment." (Madden,J.,1973).

Alternatives to Chronic Experiments

It is frankly not easy to find alternatives to these animal investigations of chronic surgical conditions, yet the suffering connected with the above experiments, and with burn experiments, seems most urgently to demand it.

Lacking suggestions for totally satisfactory substitutes, a piecemeal approach may be the best that can be offered at present.

1. The focus of investigation should be on the human victims of these accidents. Is this research developed and financially supported as much as possible?

2. When deaths occur, much may be learned from human autopsies. It was found experimentally that the lye stricture was caused by contraction of the full-thickness wound, rather than by pressure of scar tissue. Could this not have been deduced from macroscopic and microscopic examination of autopsy material?

3. If animals continue to be used, more attention to post-operative pain relief is imperative.

4. SURGICAL EXPERIMENTS CRITICIZED

Self-criticism in the Research Community

Animal surgery in America, and especially the lack of post-operative care, has its critics, even though the surgeon is a generally respected figure - but one who seems to flash like a meteor through the hospital on his way to and from important work in the operating theatre. No doubt he follows up on his own human cases with meticulous care, but, because of lack of time, any animals which he has been experimenting on are not likely to see much of him after the actual operative procedures. He assumes that the laboratory technician will take care of them. Inadequacies which may result have been testified to by two veterinarians who have been in a position to know what can happen on the inside. Dr. Joseph E. Pierce, D.V.M., of the National Institutes of Health, has pointed out:

"Cruelty to laboratory animals can be represented by the misuse of animals in experimentation. Here are some examples that now exist:...the use of unhealthy animals in experiments; ...failure to have qualified technicians and professional assistance; the production of misleading information from misuse;

failing to accomplish experiments with the minimum number of animals required; inadequate postoperative treatment."

He goes on to elaborate the last point.

"How many of us have studied animals that were several days postoperative without a body temperature recording or at least a stethescopic examination? How many of us have used the presence of a heart beat or the animal's ability to wag its tail as an indication of recovery?...How many animals are hyperthermic from infection, or dehydrated from diarrhea, or anemic from parasitism while involved in study?"

These findings, he says, can result in the publication of inadequate data. He might have added that they lead to similar experiments being repeated in order to correct the mistakes of the original ones. (Schwindaman,D.,1973,p.1279).

An even stronger statement comes from T.W. Penfold, D.V.M., for 25 years Director of the Animal Care Facilities at the University of Washington. Dr. Penfold felt so keenly about the inadequacies of technique in his own institution that he wrote to the municipality objecting to a proposal to make animals from pounds available to research institutions. He says:

"It has been brought to my attention that the City Council is reviewing the possibility to issue dogs from your city animal control center to federally licensed vendors for resale to institutions for research. My experience as a veterinarian with 25 years experience as the Director of Animal Care Facilities at the University of Washington enables me to be in a position of critical judgment in this matter. During my tenure in this position some 30,000 dogs have passed through, used both in acute and chronic research purposes, and much worthwhile information and life-saving techniques have been developed....However, all this as it may be, to release animals to vendors to sell to research institutions for research is extremely inhumane....

It is the common practice of research institutions to accept a dog or cat as a healthy animal providing that animal can stand and eat. Generally no preoperative procedures are performed as in human medicine....Postoperative care is lacking in most institutions. These animals are usually left alone to roll around in cages during the recovery stage, more often than not are not given supportive fluid therapy, and follow-up of the postoperative care by the surgeon is, in general, lacking....

Not until institutions develop the necessary guidelines in establishing presurgery evaluation for candidates for chronic surgery, and a level of postoperative care that meets the evaluation as set forth by the School of Veterinary Medicine at Pullman or the Small Animal Hospital Association, and each institution has a veterinarian to oversee such policies, should any animal be released." (Penfold,T.,1974).

"Allowed to Die"

An experience at Berg Institute, New York University, in 1973, has made Dr. Penfold's comments painfully vivid to me. With another physician, I made an unannounced and unsupervised visit to the animal quarters there at about 10:00 p.m. on a Friday evening. We noticed a cage where a small, brown and white female dog, with a gaping, undressed throat wound which ran from ear to ear, was lying in a pool of blood. She appeared to be dying. We looked in vain for an attendant to notify. Finally, we telephoned the surgeon whose name was on the cage at his home in the suburbs and told him about the dog's condition. He said he had performed an experiment on the heart vessels (an anastomosis) which "had not been successful," and that the dog should be "allowed to die." He had no suggestions for terminal care or euthanasia.[1]

Although it is obvious that the dog was not being treated in accordance with the "professionally acceptable standards" required by the Animal Welfare Act, it may be unfair to condemn the Department of Surgery at Berg Institute on the basis of this one case. But what was objectionable was the obvious lack of concern.

Dr. William A. Nolen, a distinguished surgeon and author of the book *The Making of a Surgeon*, (Nolen,W.,1970), makes a

---

[1] *We made our own arrangements for immediate euthanizing of the animal through an outside source.*

pertinent comment:

"There are some professors of surger~
train a man to be a surgeon by letting hi...
don't believe these professors. Dogs aren't people, a...
matter how humane he is, a surgeon doesn't operate on a dog
with the concern he shows for a human patient." (*Ibid.*,p.159).

Nolen concedes that it is possible to learn a particular technique by operating on an animal, and by a coincidence cites the
same operation - anastomosis of the aorta - which had been performed so unsuccessfully at Berg Institute. This was a comparatively rare procedure at Triboro Hospital, where Nolen was
assistant resident, but the chief surgical resident decided to
carry it out although he had never seen one done. He practiced
on twelve dogs, each of them obtained from a pound in Queens,[1]
with Nolen reluctantly standing by. The dogs were all humanely
euthanized with an overdose of anesthetic after the operation,
but Nolen says the experience forever cured him of any desire
for a research career in a dog laboratory.

"Whack Them Hard"

Another surgeon, Yale graduate Elizabeth Morgan, discovered
that she did not like animal research after nine months' exposure to it as a medical student. In her book, *The Making of a
Woman Surgeon*, she describes her assignment to a laboratory at
Oxford University. She worked there for a specialist in infectious diseases. Her project was to study the white blood cells
of rats and mice who had been infected with the parasite causing
trichinosis and with pneumonia bacteria. The nearly moribund
animals were killed and put through a meat grinder so that the
proliferating infectious organisms could be recovered.

[1] *This was in 1958 when New York pounds were obligated to supply
animals on demand to research institutions. Since the repeal
of the Metcalf-Hatch Act in 1979, it is no longer mandatory.*

ALTERNATIVES TO PAIN

Their standard method of killing the rodents, explained one of Morgan's co-workers, was either to chop their heads off, or to hold them by the tail and whack them hard aginst the edge of the counter. Unfortunately the rats were so inbred their tails tended to come off before "you could crack the head against the tabletop."

Morgan hated watching the animals sicken, and although she resorted to ether to finish them off, she also hated to kill them. She says that luckily President Nixon began cutting research grants, and by the time she returned to America most chiefs of surgery had found that trainees could get along without experimental animal research during their residency. (Morgan, E.,1980,p.100-104).

"Publish or Perish"

Thus there are many aspects of surgical research which justly arouse the humanitarian's ire. To recapitulate: the experimental animal may be sick to begin with; it may suffer unduly because anesthesia is withheld ("would defeat the purpose of the experiment"), or is inadequate, or is masked by curariform drugs; postoperative pain relief may be omitted, and other postoperative care may be so negligent as to add to the suffering; the animal may be subjected to multiple procedures; and, throughout, may be treated with a lack of concern which compounds its distress and fear (particularly if it is a former pet rather than a laboratory-bred animal). Finally, it may be killed inhumanely.

Many of the above deficiencies will introduce variables into the experiment and may account for the frequency with which the same procedures in the hands of two groups of investigators are often at odds. To sort this out more experiments have to be done and more animals sacrificed.

Worse than practice surgery (which may bring *some* improvement in technique) or "basic research" (with luck, *some* practical application may emerge), is research which is done merely to make a splash in a journal or to gain position and prestige in a research-oriented hospital center. Dr. Nolen, describing a young surgeon, "Al," who was doing research for just those reasons, maintains that "unfortunately, for every single dedicated, capable research surgeon there are dozens like Al. They clutter up the literature, burn up government and foundation funds that could be spent for a better purpose and waste their own talents, energy and time." (Nolen,W.,1970,p.14).

Reform is obviously needed but where will it originate? Governmental regulation has proved ineffective, and editors of scientific journals rarely refrain from publishing research, however superfluous, if among the co-authors is one with a prestigious name. Perhaps a judicious redistribution of grants, stimulated by citizen action and political pressure, is a possibility. And it would help if the American College of Surgeons or similar professional body were to emulate the position taken by their British colleagues in the Royal College of Surgeons. As described by Professor D.E.M. Taylor, representing the Royal College at a symposium of scientists and animal welfare workers where "The Welfare of Laboratory Animals" was the subject:

"The College is actively encouraging the development of alternative methods of investigation, such as tissue culture, *in vitro* modelling, clinical assay procedure and computer simulation and recommended that 'before embarking on a project entailing the use of animals, a research worker should satisfy himself that no alternative technique will meet the needs of his investigation.' However, there are still areas where there is as yet no reliable alternative to animal experimentation if the art and science of surgery is to continue to advance; these include the production of antisera, the assessment of toxicity and teratogenicity of new implant materials and much work in surgical physiology. We are all agreed that where animal experimentation is essential the standard of care and attention to our animals while they are in our charge should be based on the criteria that we would consider necessary for our patients. This regard for experimental animals is re-inforced by a knowledge that unless the highest standards of care are exercised the results of the investigation may be invalid." (Taylor,D.,1977,p.103).

Ten

TERATOGEN TESTING

The Tragedy of Thalidomide:
Misleading Animal Tests

The thalidomide tragedy focused the attention of the public on the inadequacy - prior to the 1960's - of the testing of drugs, food additives and environmental chemicals to see if these substances were "teratogens," that is, could cause malformations in the offspring of women who had been exposed to them. DCBL, the pharmaceutical branch of the large liquor company, Distillers, which promoted thalidomide in England, relied on the test reports, later shown to be entirely inadequate, of the German firm, Chemie Gruenenthal, which developed the drug in the mid 1950's. The final toll of this failure to recognize that thalidomide was a teratogenic agent was thousands of deformed children, most of them born with rudimentary or missing limbs. Their mothers had received the medication to control nervousness and nausea during pregnancy.

Tests of thalidomide on many pregnant animals had failed to produce deformities in their offspring; only after the drug had been in use for some years, and a particular strain of rabbit had been tested between the 8th and 16th day of pregnancy, did malformations show up in the litters. This happened during the phase of "organogenesis," when the body organs begin their development - a period of maximum sensitivity to teratogenic agents.

The significance of this, and of the clinical observation
that in humans too thalidomide was teratogenic only if given
during a limited period in pregnancy, was demonstrated by the
1971 experiment of Lash and Saxén. In tests on cultured human
embryonic tissue it was found that in the presence of thalidomide there was a marked decrease in cartilage development of
tissues of embryos five to seven weeks old, predicting the limb
deformities which actually occurred in the children of women who
had received the drug during the second month of pregnancy.
(Lash,J.,1971).

If tests similar to this had been performed at the time the
drug was first developed, instead of misleading animal experiments, the tragedies which resulted from its use might have been
avoided. Ironically, the developer of thalidomide, Chemie
Gruenenthal, was acquitted in 1970 after a long trial because
many eminent medical authorities testified that standard animal
tests could never be conclusive for human beings. (Ruesch,H.,
1978,p.361).

Many Variables Interfere with Animal-Human Comparisons
The factors which modify test procedures in animals, and
hence the variables which can affect comparison (extrapolation)
between these and the results in man, are virtually endless.
Here are some of them. (Grice,H.,1975,p.135).

1. Anatomical differences (one placenta in humans, two in
common test animals like rodents and rabbits).

2. Difference in metabolic patterns between animals and humans both on the adult and fetal level, affecting the concentration, distribution and excretion of the test substance.

3. Variations in response of animal species; e.g., rats are
relatively insensitive to cortisone and thalidomide; the cat

exhibits unique pharmacologic reactions. "Spontaneous" malformations unrelated to known teratogens appear in the offspring of all species, including man.

4. Short gestation period in most experimental animals compared to man. Thus tissue levels of chemicals under test may not reach levels comparable to that producing teratogenicity in human pregnancy.

5. Sensitivity of animals to many environmental factors which may cause variable results. For example, pesticide residues in bedding can affect metabolism; temperature, high or low barometric pressure and audiovisual stimulation can induce congenital malformations in some species; unbalanced diet, including vitamin excess or deficiency, can be teratogenic, while inconsiderate treatment, the number of animals per cage, the age of the mother, and the season of the year have all been shown to influence the test results.

6. Route of administration of the test substance; vehicle or solvent in which chemicals are given; dosage. For instance, large amounts of test material can affect palatability of food and water and produce nutritional imbalance. An experiment to test Coxsackievirus B3 produced severe fetal growth retardation. But this proved to be largely the result of destruction of the mother's pancreas and consequent secondary undernutrition of the fetus rather than a direct effect of the virus on the fetus. (Coid,C.,1978,p.675).

7. Potentiation - when a compound not teratogenic in a species may yet boost up another compound from inactivity to active teratogenicity. Or a physical state may be a potentiator: immobilization, normally not teratogenic in rats, potentiates Vitamin A teratogenicity. It is even conceivable that a hitherto unknown teratogen might potentiate a known one, or two unknown might act together to produce teratogenicity and thus

explain presently unaccountable human malformations. The Grice report of the Canadian Ministry of Health comments:

"The problem of the requirements for tests on combinations of chemicals has created concern in regulatory agencies. It is recognized that to test, in combination, the multiplicity of compounds to which man is exposed daily, is totally impractical." (Grice,H.,1975,p.153).

A few pages later, the report half admits the near-futility of the search for assurance against teratogenic risk with the statement:

"While predicting hazards it is necessary to make an allowance for the variability of species and individuals and other unknown factors which might adversely influence the teratogenic consequences and whose significance is difficult to evaluate. A subjective compensation or a 'safety factor' must, therefore, be applied." (Ibid.,p.160).

A "subjective" approach to "variability" and "other unknown factors...difficult to evaluate" may have been an honest description of teratogenicity research up to 1975, but today it is fair to ask, "Are these criteria reconcilable with the present 'state of the art'?"

In his book about alternatives to animal experiments D.H. Smyth says: "If there is one field of investigation where the use of animals is justified it is the avoidance of malformed babies." (Smyth,D.,1978). He couples this with considerable scepticism of in vitro methods in this area. Everyone would agree in principle that no efforts should be spared to avoid the terrible tragedy of such births, but conventional animal experimentation has not been strikingly effective in achieving this. The numerous variables which make teratogenicity experiments hard to interpret and to extrapolate to humans are probably responsible. Therefore, as happens so often, there are scientific as well as humanitarian reasons for exploring other than traditional methods of investigation.

Alternative: Epidemiological Studies

Since we are concerned primarily with effects on the human species, the first area to consider is that of studies on human beings. For example, it has been found that the nicotine and carbon monoxide which enter the system of women who smoke when pregnant can retard fetal growth, produce lower than normal birth weights, and increase still-births and infant mortality. (Am.Cancer Soc.,1978). Other epidemiological studies on the drug and eating or drinking habits of childbearing women can be correlated with teratogenic effects. Clinicians, especially obstetricians and pediatricians, should be encouraged to report suspected adverse reactions of drugs in confidence to appropriate regulatory bodies. (Inman,W.,1971). In Britain, such reports are made to the government's Committee on Safety of Medicines.

Alternative: Human Fetal Research

The above studies are valuable, but knowledge they bring may be difficult to relate to specific birth defects. An earlier line of defense lies in the use of fetal material or the fetus itself in research. The thalidomide study of Lash and Saxén using embryonic tissue cultures, cited above, is a case in point. This, of course, is a sensitive subject, but some useful guidelines have been proposed by an advisory group appointed in 1970 by the British Government "to consider the ethical, medical, social and legal implications of using fetuses and fetal material for research." (UK DHSS,1972).

Their principal recommendations were as follows:

1. Any fetus less than 20 weeks of age (weight 400-500 grams) is pre-viable and has not reached the stage at which it can exist as a separate living entity. Only fetuses of less than 300 grams should be used for research. At this weight those parts of the brain on which consciousness depends are very poorly developed structurally and show no signs of electrical activity.

2. Where a fetus is viable after separation from its mother

it is unethical to carry out any experiments on it which are
inconsistent with treatment necessary to promote its life.

3. It is unethical to adminster drugs or carry out any
procedures during pregnancy with the deliberate intent of as-
certaining the harm that they might do to the fetus.

Within these bounds, fetal research can be a valuable alternative to animal investigations. The British advisory group
includes an Appendix of 51 suggested research projects; here
are a few which could relate to teratogenicity:

1. Fetal size in relation to maternal smoking habits in
and before pregnancy.

2. Carbohydrate metabolism in hypoxic [oxygen deficient]
fetuses and the effects of maternal dextrose infusions [intra-
venous feeding].

3. Vitamin A content and activity of liver (and brain).
(See Kochhar experiment below, p.181, for teratogenicity of
Vitamin A).

4. Culture of renal tissues to elucidate the development
of fetal renal malignancies.

5. Alterations in trace metal metabolism in relation to
protein and electrolyte levels in amniotic fluid.

6. Placental metabolism: studies of the transfer of drugs,
bacteria, and biochemical substances.

7. Chromosome studies, including abnormalities found in
therapeutic as well as spontaneous abortions.

The advisory group adds this comment:

"There is a particular need to determine the ability or
otherwise of the fetus to deal with substances, including drugs
given therapeutically to benefit the mother, which may cross
the placenta. Observations on the pre-viable fetus are necessarily limited to a period of two or three hours. They have, however, already contributed significantly to our understanding of
vital physiological and biochemical processes before birth on
which the development of a fetus into a normal child essentially
depends...and promise to be the most hopeful approach to understanding certain failures of the human brain to develop properly
and the influence such factors as variants in sexual differentiation *in utero* may have on inherent behavioral patterns after
birth."

Human Organ Cultures

Since such a fetus in its entirety is only viable, after removal from the mother, for two or three hours, it is fortunate that its parts can be kept alive in culture. These include cells, tissues (e.g. renal tissue, as in 4 above) and even organs. The value of organ culture is illustrated by an experiment performed by Karkinen-Jääskeläinen and Saxén at the University of Helsinki. These investigators wished to discover why and how rubella infection in the mother, when it occurs during the first trimester of pregnancy, often results in the occurrence of congenital cataract in the child. "To study the problem more closely," they state, "we looked for a model and immediately ran into the first difficulty, not uncommon in research on teratology: there was no good animal model for the study." However they found that they could produce cataracts in the lens fibres of chick embryo by infecting them through the eggshell with a viral disease similar to rubella, namely mumps. They observed that the chick lens, which forms from an invagination of the surrounding cells which then becomes a sealed vesicle, can only be invaded by the virus as long as it remains in open communication with the exterior. They continue:

"We now returned to the original problem of human congenital cataract and collected young embryos from therapeutic abortions. One of the eyes of each human embryo was infected with rubella virus *in vitro* and its pair, the uninfected control, was cultured in otherwise identical conditions. Twelve embryos obtained when less than 5-6 weeks old, had open-stage lens vesicles, and some hundred embryos were at the closed stage."

However, 50 of the latter had their lens vesicles opened surgically before infection and these, as well as the 12 whose vesicles were naturally open, developed cataractous changes, whereas the other with closed, unoperated vesicles, did not. (Karkinen-Jääskeläinen,M.,1976,p.275).

Advantages and Disadvantages of Organ Culture

The above paper is entitled "Advantages of organ culture techniques in teratology." The authors discuss some of the advantages and disadvantages of organ culture. Among the advantages are the following: it can be studied free of maternal or placental confounding factors; its stage of development can be exactly timed by observation; exact dosage and exposure time to the teratogen is known and can be simply controlled by changing the culture medium. Disadvantages are the necessity of working with very small organ fragments because of nutritional limitations - and even these fragments cannot survive for long; the laboriousness and delicacy of the technique as compared with simpler cell culturing; and the absence of the natural conditions which are found in whole animals experiments: for instance in situations where maternal metabolites may be more toxic to the fetus than the substance acting directly on it without maternal mediation.

Four New Alternatives in Teratogen Testing

1. Enzyme Activation

That this last disadvantage can be overcome to some extent is illustrated by recent work of Manson and Simons who tested the effect of the teratogen cyclophosphamide on an organ culture of mouse embryonic limb buds. It is known that the teratogenic effect of this drug is not demonstrable unless it is first activated by an enzyme system known as "mixed function oxygenases" (MFO). Since hamster embryonic cells have high MFO activity, they were co-incubated with the limb buds and cyclophosphamide, whereupon the metabolites of the drug induced abnormal limb development (a similar effect occurs in humans). (Manson,J.1979).

2. A Battery of Tests to Assess Drug Action

It will be recalled that the thalidomide tragedy occurred in spite of the drug's having been tested on thousands of pregnant animals. Since the dire consequences of its use have become known, the testing of new drugs for teratogenicity on such animals has vastly increased. Desirable as it would be to reduce this often irrelevant sacrifice of animals in favor of techniques using human materials, it is not going to be possible in the near future to substitute the latter for much of the animal work. However, it might be possible greatly to reduce the numbers of adult animals sacrificed, and the duplication of tests in different species, if attention is focused on the actual mechanism of drug action, and the principles of abnormal development, through organ culture techniques and microscopic observation of the behavior of cells in culture, instead of merely noting gross pathology (or its absence) in the fetus.

D.M. Kochhar has attempted this and has described it in a paper entitled, "Elucidation of mechanisms underlying experimental mammalian teratogenesis through a combination of whole embryo, organ culture, and cell culture methods." (Kochhar,D., 1976,p.485). He studied the effect of two substances, retinoic acid (Vitamin A) and 5-bromodeoxyuridine, on cartilage development in the limbs of mice embryos. Through the methods referred to in his title he was able to demonstrate various teratogenic effects, including a progressive decrease in cartilage development and disturbances in cell migratory rates as recorded by time-lapse microphotography of mouse limb bud cell cultures. He suggests that no single system but rather a combination such as he has employed may be required to assess the teratogenicity of a new drug before it can be confidently released on the market.

3. Biosynthetic Patterns of Cells

Since the number of potential teratogens has vastly out-
stripped the possibility of testing them by laborious methods
using the intact animal, rapid *in vitro* assay systems must be
found. One such has been developed by Ruth Clayton of the Institute on Animal Genetics at the University of Edinburgh. She
first described it in a paper published in 1976 (Clayton,R.,
1976,p.473), and discussed recent developments in a FRAME symposium at the Royal Society in 1978. FRAME's abstract follows:

"The system developed in their laboratory is based on the
premise that there is a relationship between the synthetic
capacities of cells and abnormal development. All the major
protein products of the cell are analysed in microgels where
abnormal synthetic patterns can be readily identified by auto-
radiographic and densitometric techniques. The method has the
added advantage that it can be adapted for routine use. Samples
of amniotic fluid taken from treated and untreated animals are
applied to foetal cells (e.g. lens, neural retina, kidney & limb
fibroblasts) in culture and then the effects on biosynthetic
patterns are recorded. Using this system, they were able not
on.y to predict correctly which of two unknown substances was
was teratogenic, but also to identify the effects of the teratogen." (Anon.,1978b).

4. Inhibition of Cell Attachment to Lectin-treated Surfaces

A still more rapid screening system for teratogens has been
described by Braun and Nichinson. Tumor cells attach rapidly
and irreversibly to plastic surfaces treated with plant lectins.
Teratogens inhibit this attachment. The investigators tested a
series of agents. Thirteen did not inhibit cell attachment and
were probably non-teratogens. Another nine were also non-
inhibitors but were considered to be false negatives; i.e., they
were thought to require metabolic activation for their teratogenicity to be expressed. Twenty-five other known teratogens
inhibited cell attachment. (Braun,A.,1979).

Although some of the work described above is exploratory and requires further confirmation, it is obvious that there are now many systems being developed in teratogenicity testing which are approaching closer to the use of human material and are reducing reliance on the old-fashioned intact animal experiment. The pessimistic attitude to *in vitro* work in this area expressed both by D.H. Smyth (Smyth,D.,1978) and by H. Grice and his collaborators (Grice,H.,1975) is no longer justified.

Eleven

TESTING BIOLOGICAL PRODUCTS

1. HORMONES

Hormones are chemical substances produced in organs of the body, chiefly but not exclusively from the endocrine or ductless glands. They move throughout the body in the blood, regulating functions such as metabolism, growth, pregnancy and sexual development. They include insulin, thyroid hormones, and the gonadotropic, androgenic and estrogenic sex hormones among many others.

Alternatives

Tests to measure the amount of hormones in body fluids which formerly used animals have now been partly superseded by biochemical *in vitro* methods.

1. Instead of rabbits, mice and toads once used in great numbers in the gonadotropin pregnancy test, pregnancy may now be diagnosed by an agglutination test which eliminates the animals. (*PDR*,1975,p.2028).

2. Instead of the rats, mice and birds formerly required to detect estrogens, progesterone, corticosteroids and 17-oxosteroids in the body fluids, a chemical assay is now the method of choice. (NRC,1971,p.134).

Insulin

Although some hormones like the steroids are produced by chemical synthesis, the most important sources of insulin and

184

certain other hormones are the endocrine glands of animals
slaughtered for meat. The insulin assay used by the Drug Bioanalysis Branch of the U.S. Food and Drug Administration is
described in the *U.S. Pharmacopeia XIX*, p.611 (1975). Insulin
manufacturers are required to follow the same method. The so-
lutions to be tested are compared with official standard insulin: different dilutions are injected into rabbits and after
one hour, and 2 1/2 hours, the effect on the rabbits' blood
sugar is compared. This gives a quantitative evaluation of the
potency of the test substances. I asked Dr. John Collins of
the Drug Bioanalysis Branch whether the rabbits were anesthetized for this procedure and he replied that the pharmacopeia
does not call for it:

"My opinion is that this is impractical for a number of
reasons. Aside from the possible physiological intrusion to
insulin response caused by anesthesia one must consider that
practicality demands that the rabbits be used more than once,
and it is certainly undesirable to expose animals to chronic
anesthesia. The rabbit test is not usually hard on the animal;
indeed an outpouring of epinephrine[1] in the course of the assay
would result in glucose elevation and possible assay invalidity.
The rabbits do receive a subcutaneous injection through a sharp
needle, and never appear to show any distress. If the blood
sample is obtained carefully from a marginal ear vein, the animal also seems to suffer little. In my opinion, anesthesia is
undesirable in this case because it is comparable to not using
anesthesia for penicillin injection or venipuncture in the
human." (Collins,J.,1980).

He said that the rabbit bioassay had replaced the "mouse
convulsion" test in the U.S. many years ago; however, the
latter is still done in England using vast numbers of mice.
Even though the injected mice are given glucose to bring them
back to normal as soon as the convulsion begins, a convulsion
is nevertheless a required end-point of the test; thus it is

1
 If the rabbits were in fact experiencing pain or fear.

an "all-or-none" procedure which does not yield quantitative
information. Dr. Collins added that the Identification Test A
on p.253 of the *U.S. Pharmacopeia XIX,* which describes an injection of test insulin into six rabbits, three of which must
convulse and recover after glucose administration, is not used
by the Food and Drug Administration. If it is obsolete, it's
a pity the pharmacopeia continues to publish it.

A blood glucose test on mice has now appeared in Appendix
14 of the *British Pharmacopeia* as an alternative to the mouse
convulsion test. P.A. Herman of the British Home Office Inspectorate writes:

"You will be interested to know that it has been estimated
that the introduction of the blood glucose method of assay into
the Addendum of the *British Pharmacopeia* may reduce the number
of experiments in this area quite considerably. Indeed the
estimate of one establishment carrying out this work is that
the number of experiments it performs will be reduced by over
75%." (Herman,P.,1979).

Dr. Collins says that his laboratory is looking into this
test as a potential replacement for the rabbit method.

Synthetic Insulin and Other Hormones

The chemical synthesis of bovine insulin was accomplished
in the late 1960's, and in the early 1970's human insulin was
synthesized by the Swiss drug firm Ciba-Geigy. The quantities
produced were apparently small, so the announcement in 1978 that
the San Francisco pharmaceutical firm Genentech had synthesized
human insulin, using a bacterial "gene-splicing" technique which
has a potential for producing significant and eventually commercial amounts, earned headlines for the company and for their
collaborators at the City of Hope Medical Center at Duarte,
California. A further technical advance, the bacterial production of human proinsulin (the precursor of insulin), was
announced by Genentech in Mar. 1980. (Anon.,1980g).

Genentech had previously been associated with the synthesis
of the brain hormone somatostatin, and in the summer of 1979
they had achieved the synthesis (almost simultaneously with
another team at the University of California, San Francisco) of
human growth hormone - all through the use of recombinant DNA
technology, or gene splicing. The major producer of insulin
from animal sources, Eli Lilly, is also in the race to make
synthetic insulin in commercial quantities. With 1,300,000
American users of insulin, the financial stakes are high.

Although the coming of synthetic insulin is a big step
along the road toward the elimination of animal in favor of
chemical testing, it should be mentioned that no attempt to
synthesize the protamine modifier of insulin has been reported.
Protamine is an amino acid-rich substance obtained from fish
sperm (now mostly imported from Japan); it makes insulin long-
lasting and is prescribed for some diabetics. Animals are
needed to determine the biological reaction curve owing to
the variables from the differences in lots of this insulin
modifier. However, there is an alternative: an extended insulin
zinc suspension known as "Ultralente." This is as long-lasting
as protamine insulin but has a modifer which is inorganic and
thus does not require the animal assay.

Hoped-for Phasing-out of Testing with Animals

As for insulin itself, and the expected replacement of the
beef or pork hormone with a product derived from human genes by
DNA recombinant technique, Dr. Collins makes the following encouraging comment: "If the product is as good as we all hope, it
is quite possible that a few chemical methods such as electro-
phoresis or chromatography can suffice to identify it, and
animals will not be needed." (Collins,J.,1980).

And if the product were always made in exactly the same manner,[1]
each batch would be identical and animal testing for potency
and efficacy would also be superfluous.

2. VACCINES AND ANTITOXINS

This section has to do with the viruses, minute living
parasites which inhabit living cells; with bacteria, and the
toxins which some of these produce; and with the body's weapons
against them: antibodies and antitoxins. These defenses are
mobilized in answer to a call to arms, and this is sounded when
information comes from a surface or coat protein carried by the
microorganisms in the form of a code. Certain specialized body
cells, especially the lymphocytes, are able to "read" and react
to this code, which consists of a particular sequence of protein molecules formed on the surface of the microorganism out
of the DNA - the genetic material of living things. This surface protein, when it provokes a cell to produce antibodies and
antitoxins, is called an antigen.

When on the warpath, we can imagine that these microorganisms, like placard-carrying militants, display codes spelling
out messages such as "Death to the cells!" This may result in
an orderly counterattack by the lymphocytes and other members
of the immunological defense system. Or it may panic the cells
into an overreaction, like the body's response to gram-negative
bacteria which display lipopolysaccharide in their walls and
which our tissues read, in Lewis Thomas's words,

"...as the very worst of bad news. When we sense lipopolysaccharide, we are likely to turn on every defense at our

---

[1]
*The U.S. Food and Drug Administration's "Good Laboratory Prac-
tice Regulations" (US FDA,1978) are intended to assure the
high quality of testing for safety of products such as these.*

disposal; we will bomb, defoliate, blockade, seal off, and
destroy all the tissues in the area....Pyrogen is released from
the leukocytes, adding fever to hemorrhage, necrosis and shock.
It is a shambles." (Thomas,L.,1974,p.92).

These overreactions are part, and sometimes the most uncomfortable and dangerous part, of what we call "disease."

However, we have learned how to use the milder forms of
immunological defense to our advantage by actually cultivating
the virus, then "taming" it either by weakening (attenuating)
or killing (inactivating) it and thus producing a vaccine. The
attenuated virus, taken by mouth, multiplies in the intestinal
tract: the inactivated virus does not multiply, so enough has
to be given by a series of injections. Both kinds of vaccine
stimulate sufficient antibodies to protect against the next in-
vasion of a virulent virus, but in their attenuated or inactivated form do not themselves produce the damaging effects of
the disease. Bacterial toxins can also be cultivated - from
the bacteria grown in the laboratory. The toxin is treated
with formalin to render it harmless, but it retains the surface antigen and can stimulate the production of antitoxin if
injected into an animal, such as a horse.

Moderating the Rigors of Tetanus Testing

If the disease is actually present, however, there may not
be time to wait for antibodies and antitoxin to mobilize; in
that case the alternative is to obtain them from a person who
has had the disease and in whose serum they are still circu-
lating. Anti-tetanus serum may be obtained in this manner.
As D.H. Smyth points out, the antibodies prepared from human
blood require "less safety testing than blood from a different
species, and the tests for the normal hazards of human blood,
syphilis and hepatitis, do not require live animals."

Tetanus toxoid production, on the other hand, involves

SOURCE	MONKEY	DOG	RABBIT	GUINEA PIG	MOUSE	CALF, SHEEP
VACCINE						
Adenovirus	P					
Canine Distemper		P				
Diphtheria Toxoid				A		
Measles	P	P				
Polio	P,A					
Rabies			P			
Rubella	P	P	P			
Smallpox			A			P
Tetanus Toxoid					A	
Whooping Cough					A	
Yellow Fever					A	

Table 1: Vaccines Grown and Assayed on Primary Cultures from Killed Animals

"P": production; "A": assay. Left: Disease vaccines; Top: Animals used in production or assay.

the painful participation of guinea pigs under U.S. Dept. of Agriculture regulations. The virulent toxin has to be tested in control animals: "For a satisfactory test, the controls must die with clinical signs of tetanus:...increased muscle tonus, curvature of the spine,...generalized spastic paralysis...." (US Laws,Stats.,etc.,1979,*CFR*,Title 9).

But Smyth notes that the (British) National Institute for Biological Standards and Control does not require that mice on whom the still active toxin is tested have to be observed until death (an agonizing one owing to the tetanic cramps). A smaller dose which merely causes a paralysis of the hind leg is sufficient for the test. (Smyth,D.,1975,p.83).

Production and Assay
of Vaccines

The production and assay (testing for potency and safety) of vaccines consume the lives of many animals, often in painful ways. In the past, many more were sacrificed. In the early years of polio vaccine, in the late 1950's, 200,000 rhesus macaques a year were being killed for that purpose in the U.S. In the early 1950's, thousands of dogs were destroyed yearly in the production of distemper vaccine. *Table 1* lists the vaccines for different diseases and the animals from which they were, or in some cases still are, derived. Most of them were grown on primary kidney tissue cultures, "primary" meaning that the cells are taken directly from the kidney of the monkey, who is killed in the process.

In recent years there has been a great reduction in the use of primary cultures in production, thus sparing many animals. This is because all the vaccines listed in *Table 1*, and others as well, can be grown on embryonated duck or chicken eggs, or on a human diploid cultured cell strain WI-38.

SOURCE	HUMAN DIPLOID CELL STRAIN WI-38	EMBRYONATED DUCK OR CHICKEN EGG	SEROLOGICAL FLOCCULATION TEST
VACCINE			
Adenovirus	P^1, A^2		
Canine Distemper		P^3	
Diphtheria Toxoid			A^4
Measles	P^1	P^2	
Polio	P^2		
Rabies	P^5, A^5	P^5	
Rubella	P^1	P^2	
Smallpox	P^1	P, A^2	
Tetanus Toxoid			A^4
Whooping Cough			A^6
Yellow Fever	P^1, A		

Table 2: Vaccine Grown and Assayed by Methods
Which Avoid Animal Killing

"P": production; "A": assay. Left: Disease vaccines; Top: Substances used in production or assay. References: 1. Hayflick,
L.,1970; 2. US Laws,Stats.,etc.,1978,*CFR*,Title 21,Sec.170-299,
3. Russell,W.,1959,p.83; 4. Keele,C.,1962,p.311; 5. Tint,H.,
1974; 6. *PDR*,1978,p.98.

The last named was developed at Wistar Institute by Dr. Leonard
Hayflick. (Hayflick,L.,1961). The WI-38 strain was obtained
from embryonic lung tissue and, unlike a primary cell culture,
can be propagated for many generations before dying. There is
no risk that the cells might harbor a virus from the animal
donor, nor do they undergo cancerous transformation. They can
be produced in large quantities and can be preserved by freezing.

A rabies vaccine has been grown on the WI-38 strain, which
unlike the presently used vaccines, is painless and is free from
the complication of postvaccinial encephalomyelitis. It has been
successfully tested on over 1000 people since 1975, and now
awaits FDA licensing. (Anon.,1980e).

Table 2 demonstrates the saving of animals which is now
possible through the use of alternative production or assay material. Of course these alternatives were developed, not for
humane reasons, but because they were more economical and safer.
Probably they are not fully utilized even now. Much painful
testing of the vaccines on animals is still required by government agencies, although assay methods using cell cultures are
available. A plaque-test is said to be four times as sensitive
as one using mice (Tint,H.,1974), and instead of the old way of
testing Yellow Fever virus by injecting it into the mouse's brain
and observing the incidence of paralysis, the World Health Organization now recommends a technique employing plaque formation in
a monkey kidney cell culture. (Whitaker,A.,1977,p.28).

The Suffering of Unprotected
"Control" Animals

In spite of the gains that have been made, especially the
changeover to embryonated eggs (from vaccine production on primary cell culture requiring the death of the animal donor)
some animals still die painfully to guarantee the purity of

biological products. The ones that suffer most are not those
that are injected with the vaccine to be sure it contains no
bacterial or viral contaminants (they might fall ill, of course,
if there were a contaminant and they were infected - but that
presumably is rare). The real martyrs are the mice and guinea
pigs who do not receive the protective vaccine against diseases
such as cholera, anthrax, typhoid and tetanus, but, along with
a matched group of vaccinated animals; are inoculated with the
disease-causing bacteria. Thus Bayvet Corporation, Shawnee
Mission, Kansas, reporting to the USDA under the Animal Welfare
Act, described pain or distress inflicted on 9,448 guinea pigs,
4,850 hamsters and 1,239 rabbits - "without pain-relieving drugs."
Their explanation:

"These animals were utilized to test the efficacy of biological products, using virulent challenge material, by the
procedures and methods outlined in the Federal Register, Title
9, *CFR*, 113-117. Using drugs in these animals would invalidate
the test and render it useless." (USDA/APHIS,1976,Bayvet Corp.).

In a "successful" test, proving the efficacy of a vaccine -
such as the one for tetanus toxoid described on p. 189 - the
immunized animals survive and the unprotected controls die,
painfully.

Methods of Polio Vaccine Manufacture: Monkey
Tissue Versus Human Cell Culture

The reduction in available rhesus monkeys as a result of
the ban on exports from India as well as from other countries,
calls for a reappraisal of the extravagant consumption of
primates in research, and in the manufacture and testing of
polio vaccine - this being the largest use of rhesus in the
U.S. Lederle Laboratories of Pearl River, New York, is the sole
producer in this country of the vaccine. They make the Sabin
attenuated virus under the trade name "Orimune," growing it on

primary monkey kidney tissue. Pfizer, unfortunately, has discontinued production of the polio vaccine which was grown on
human cell culture, and which at least spared the monkey kidney
tissue donors. Wellcome Laboratories in England still produces
the vaccine on human cell culture, and A.M. Whitaker, describing
the advantages of starting with human cells, has pointed out that
cell cultures prepared from monkeys are often infected by one
of the many diseases to which the simians are prone (60 monkey
viruses have been identified), so that about 55% have to be dis-
carded during vaccine manufacture. (Whitaker,A.,1977,p.25).

Polio Vaccine: Production, and Tests for Contamination and Neurovirulence

After the cell culture has been prepared and tested, the
seed virus is placed on it, is allowed to multiply, is attenuated by dilution and the vaccine fluid is "harvested." (Antibiotics in small quantities have also been added). *Table 3* out-
lines the steps of manufacture of attenuated oral polio vaccine
in Great Britain, a process similar to that used in the U.S.
(UK DHSS,1977).

The testing of the cell cultures, whether of monkey or human origin, and later of the vaccine, is carried out both in test
tubes (*"in vitro"*), and in animals (*"in vivo"*). Rabbits, guinea
pigs, suckling and adult mice are used to detect extraneous, contaminating viruses and, finally, neurovirulence is tested in monkeys. LeCornu and Rowan comment:

"It should be noted that the same tests for extraneous
agents are required whether the vaccine is produced on monkey
kidney cell cultures or human diploid cell strains. Extensive
testing is understandable in the case of monkey cell cultures
since monkeys harbor many unwanted pathogens, but the human
cells are derived from a seed stock which has been thoroughly
tested. Repeated testing is therefore wasteful of time, money
and animals."

Virus seed
(Vaccine virus)

Show seed to be free from extraneous agents,
pass monkey kidney test.

Standard method (monkey cells)	*Alternative method (human cells)*
Grow virus:	Grow virus:
75% as monkey kidney cell cultures (primary) from quarantined monkeys.	90% as human diploid cell cultures (strains) from cell bank.
25% as controls.	10% as controls.
Test for haemadsorbing viruses by addition of red blood cells.	Test for haemadsorbing viruses.
Test supernatant fluids for extraneous agents on monkey and human cell cultures	Test supernatant fluids for extraneous agents on monkey, human and rabbit cell cultures.
Test for simian herpes B virus in rabbit kidney cell cultures.	
Harvest virus.	Harvest virus.

Virus harvest, by whichever method
grown, then undergoes following tests.

Monovalent virus harvest

-Test for bacteria, fungi, mycoplasma.
-Test for extraneous viruses on monkey cell culture (and human cell
 culture for virus grown on human diploid cell culture).
-Test for simian herpes B virus on rabbit kidney cell cultures.
-Test for extraneous viruses in animals (rabbits, adult & suckling
 mice & guinea pigs).

Pool virus harvests

-Identity test.
-Measure virus concentration.
-Virus characteristics - *in vitro* virulence tests.
-Monkey neurovirulence test (35 monkeys).
-Test for bacteria, fungi & mycoplasma.

Blend monovalent vaccines - monovalent, bivalent, or trivalent
vaccines (i.e. 1, 2 or 3 types).

-Test for identity.
-Measure virus concentration (i.e. potency test).
-Sterility test.
-Abnormal toxicity.

Table 3: Production of Live, Attenuated Oral
Polio Vaccine (UK DHSS,1977).

For vaccine grown on human cell cultures, they suggest testing
large pooled batches rather than numerous small ones. (LeCornu,
A.,1928,p.10). Dr. J.C. Petricciani of the Bureau of Biologics,
Food and Drug Administration, also mentioned this as a way of
reducing animal wastage, but added that if one of the larger
batches failed the test the manufacturer would have more expense
to replace it. (Petricciani,J.,1978).

Alternatives in Polio Vaccine
Production and Testing

A number of other alternatives to animals in polio vaccine
production are in use or are being developed. Here are several
of them and the tests they might replace:

1. *Test for extraneous viruses in animals.* Unfortunately,
some of these contaminating viruses can still only be detected
in the whole animal: for example, Hepatitis A or B and most
arboviruses. For others, there are alternatives.

Alternatives: "In vitro" tests have been developed, however. Some viruses will cause added red blood cells to adhere
firmly (haemadsorption) to the cell culture which the viruses
are contaminating. Others may be detected by electron microscopy. The haemadsorption test is already included in the regulatory requirements (cf. *Table 3*); electron microscopy is used
by Wellcome Research Laboratories and, even though it is of low
sensitivity, may become part of the official requirements in the
future. (Whitaker,A.,1977,p.29).

Primary cell cultures are always associated with wastage
of animal life since they involve the killing of the animal.
Although official regulations specify the use of primary monkey
kidney cell cultures in testing vaccines for contaminating viruses, A.M. Whitaker of Wellcome states that monkey kidney cell
lines (cells which have been sub-cultured, proliferate and go

through numerous divisions) are equally sensitive to these contaminants. (*Ibid.*,p.28). The use of such lines would save many
valuable primate lives.

2. *Tumorigenicity Test.* Cells from the culture are injected into immunodeficient mice to see whether tumors are
induced. If they are, the culture has lost its normal character and contains transformed (cancer) cells.

Alternatives: There are several short-term tests for
tumorigenicity which might be substituted: for example, the
ability of tumorgenic cells to grow in semi-solid medium;
or the Ames *Salmonella* test (cf.p.87).

3. *Neurovirulence Test.* This is done to confirm that the
vaccine virus has not reverted to virulence. LeCornu and Rowan
(LeCornu,A.,1978,p.11) have this comment:

"The neurovirulence test...compares the pathogenicity of
successive polio vaccine batches to the original seed vaccine
virus....Basically, it involves the injection of the vaccine
and reference virus intraspinally and intracerebrally into
[two groups of] at least 25 and 10 susceptible monkeys respectively."

In 1955, a year after the first mass trial of the Salk vaccine (containing inactivated virus in contrast to the Sabin attenuated virus) a large number of people contracted polio because
of a technical fault in the inactivation process. Before this
incident, vaccine safety had been assayed by a tissue culture
test which had a greater margin of safety than the intracerebral
injection of monkeys which was also being used. However, after
1955 it was decided to abandon the tissue culture test for safety
in favor of more extensive monkey tests. LeCornu and Rowan point
out that "this appears to be an instance where too much time has
been spent trying to improve an unsatisfactory animal test when
more effort to standardize and improve the tissue culture test
would have been more productive." (*Ibid.*,p.6).

As it is, the safety tests on monkeys, performed by the
manufacturing company, are then repeated by the control authorities (in the U.S., the F.D.A.'s Bureau of Biologics). American
regulations require 30 rather than 10 monkeys to be tested intracerebrally. Since all monkeys have to be killed after a certain
number of days to allow tissue examination for evidence of polio,
140 are sacrificed for each batch of vaccine in Great Britain
and 220 in the U.S.

Reduction of animal wastage: LeCornu and Rowan state:
"Ideally, duplicate testing should be unnecessary and the practice needs to be reviewed. It certainly offers one opportunity
of reducing the number of monkeys needed." The problem is to
develop a very stable seed virus, in which case "it should only
be necessary to carry out neurovirulence tests on the seed, as
is now done with rubella and measles live attenuated vaccine."
(*Ibid.*,p.11-12). At present, the Sabin seed strains are unpredictable and continual passaging does not automatically
decrease neurovirulence.

Other in vitro tests for neurovirulence: Two of these, the
"d" and "t marker" tests, cannot be called alternatives because they are currently used only to complement the monkey
tests. The "d test" detects the reduced ability of avirulent
virus to replicate in the presence of sodium bicarbonate. The
"t marker test" depends on the increased temperature sensitivity
of the attenuated virus. A third test is in the experimental
stage and is being investigated at Britain's National Institute
for Biological Standards and Control. It relies on the differences in the protein bands of virulent and avirulent strains of
polio virus, which can be identified through the use of isoelectric focussing. (Magrath,D.,1978).

Synthetic Vaccines

Each lot of polio vaccine is subjected to about 29 separate
tests in order to be certain that the product is pure, potent,
avirulent and of the required dosage. But since living cells are
involved, and since these have to be grown in a medium which
sustains life and promotes growth, some of the testing is to
guard against this life growing too exuberantly (escape from
attenuation), or abnormally (mutagenicity), and the rest of the
process is to prevent other life - bacteria, extraneous viruses - from getting into the act. How much simpler it would
be, and happier for the animals now sacrificed along the way,
if these living elements could be banished and a completely pure
synthetic vaccine could be created by the chemist.

It has been done: a totally synthetic vaccine against a
natural virus has been produced by Dr. Michael Sela, of the
Weizmann Institute in Israel, and was reported at an international conference held at the National Institutes of Health in
Bethesda, Maryland, in Feb. 1979. According to an article in
The New York Times (Feb. 27, 1979), the new product was a vaccine
containing a synthetic copy of part of a virus linked chemically
to a synthetic carrier molecule and to a synthetic "adjuvant"
molecule which has the property of amplifying the recipient's
response to the vaccine. "Injected in laboratory animals, it
immunized them against the natural virus it was designed to re-
sist. The vaccine itself is of no practical usefulness because
it is directed against a type of virus called a bacteriophage,
which infects bacteria, but not humans or animals." (Schmeck,M.,
1979).

Of course this research is only the first step toward the
development of other vaccines. Diseases like malaria, and other
parasitic diseases against which there is presently no effective

vaccine, as well as the major venereal diseases, will no doubt
have priority in the development of these synthetic substances,
although one against Hepatitis B is already in the early stages
of synthesis (cf.p.249). Thus it seems likely that a vaccine
against polio will also be synthesized. Such a substance would
be standardized to a high degree of uniformity and purity, reducing the present painful "29 steps" of manufacture and assay
to a minimum.

Twelve

TOXICITY TEST PROCEDURES FOR
CHEMICAL SUBSTANCES

1. SUBSTANCES TO BE TESTED AND NUMBER
OF ANIMALS USED

The previous chapter considered the production and testing of "biologicals" - substances which are derived from living
matter. The great majority of substances mentioned in this and
the following chapter are not organic in origin but are manufactured by the chemist. Drugs, including analgesics; cosmetics;
household products; and substances used in agriculture (e.g.
pesticides) and industry, are subject to the toxicity tests
to be discussed.

A substance is defined as toxic if, when eaten, inhaled
or absorbed through the skin, it causes by its chemical action
either damage to structure or disturbance of function, or both.
Safe and effective dosages have to be estimated for all drugs,
and safe concentrations for household products. There is also
a difference in toxicity to be calculated for the ingredients
of a drug under any and all possible circumstances and for the
drug product under reasonable conditions of use. Drug manufacturers are very concerned about such things, and so are certain government agencies such as the Food and Drug Administration (with respect to food additives, drugs, cosmetics), the
U.S. Consumer Product Safety Commission (hazardous substances:
irritants, corrosives, inflammable and "highly toxic") and the

Environmental Protection Agency (insecticides, fungicides, ro-
denticides).

In all, there are in the U.S. up to two million such substances already in distribution, with 13,000 substances listed
in the toxic substances list of the National Institute for Occupational Safety and Health and 600-1,000 new substances entering
the market every year. (Eagleton,T.,1977).

Animals are extensively used in testing all these substances
for toxicity and very large numbers are involved. According to
the Institute for the Study of Animal Problems, dealers now estimate the current annual demand for all types of research and
testing at 100,000,000 animals: 50 million mice, 20 million rats
and about 30 million other animals, including about 200,000 cats
and 450,000 dogs. (Rowan,A.,1979a,p.8). It is not known how many
of these are involved in acute and chronic toxicity testing because American official bodies like the National Research Council
have failed to produce statistics on the subject, but a British
group of scientists, the Committee for Information on Animal
Research, estimated that acute toxicity testing - including the
notorious LD/50 tests - consumed one-fifth (about one million) of
the animals used in Great Britain for experimental purposes in
1975. (CRAE,1977,p.25).

One-fifth of the estimated 100 million animals currently
used in the U.S. would amount to 20 million.

2. TEST PROCEDURES

Acute Toxicity

Acute toxicity tests are a "search for untoward reactions
at a high dose level." Research facilities, particularly pharmaceutical ones, frequently mention these tests in their Annual
Reports to the U.S. Department of Agriculture under the Animal

Welfare Act (cf.p.241). They can be found under Column D of
the Report, which is for experiments "involving pain or distress
without the use of appropriate anesthetic, analgesic or tran-
qulizer." Thus Procter and Gamble:

"Acute Oral Toxicity Studies - 48 dogs. These are product
safety tests designed to determine potential local corrosive action or systemic toxic effects after ingestion of a product.
They are a necessary part of a total oral toxicity program submitted to the FDA to insure the safety of a product prior to
marketing. It is necessary that the animals receive no other
medication since their use may interfere in the evaluation of
the test. The dogs are given a specific dosage of a product by
gavage [forced feeding by stomach tube] and are observed for four
hours after dosing and daily thereafter for two weeks. Most
products cause only transient emesis or diarrhea, although occa-
sional signs of systemic toxicity, e.g., neurologic dysfunction,
are observed." (USDA/APHIS,1976,Procter & Gamble Co.).

Bristol Laboratories:

"During 1973 approximately 30 dogs and one squirrel monkey
received acute doses of a compound which resulted in convulsions
and eventual death." (USDA/APHIS,1973,Bristol Labs.).

Tulane University School of Medicine (Laboratory of Environmental Medicine):

"241 rabbits. Skin and eye irritation test and dermal
LD/50s ["lethal dose in 50% of animals tested"] were performed
according to Federal Hazardous Substances Act procedures....We
did cause some pain and distress because we were studying the
acute skin and eye irritation potential of various chemicals.
If we were to use drugs, anesthetics, or analgesics to relieve
the pain and distress we would be defeating the very purpose of
the study." (USDA/APHIS,1976,Tulane Univ.).

Column D requires a "brief explanation" of why pain-relieving drugs were omitted. Those who drafted the Animal Welfare Act
presumably expected that this explanation would serve as a check
on painful experiments performed without anesthetics or analgesics. However, in the vast majority of cases, facilities reporting such experiments circumvent this by referring merely to
the government agency regulating the test substances.

In the 1970 amendments to the Animal Welfare Act the Secretary of DHEW is told to promulgate standards for the use of pain-relieving drugs. This he has failed to do, so buck-passing references to a whole potpourri of government regulations can be used to justify the suffering inflicted on many thousands of animals.

As a reminder of what the animals have to undergo, here is a partial list of the common signs of toxicity which may be observed in acute toxicity tests in rodents: behavior: unusual vocalization, restlessness; twitch, tremor, paralysis, convulsions; muscle tone: rigidity, flaccidity; salivation, lachrymation; difficulty in breathing; diarrhea, constipation, flatulence; swelling of the sex organs and breasts; skin eruptions; discharge, including hemorrhage, from the eye, nose or mouth; abnormal posture, emaciation. These are all signs of distress which can be observed; what the animals feel can only be imagined.

The LD/50 Test

This is an acute toxicity test in which a number of animals are given different amounts of a substance to determine what dose will kill 50% of them over a predetermined period of time. If the substance is a gas, then the lethal concentration (LC/50) in air which will produce the same effect is determined.

The test is principally used to give a rough indication of the toxicity of a substance or new drug when this is unknown; or to produce, as a statutory requirement, a figure for comparison with the toxicity of known compounds. Thus the warning notices which appear on packaging under the Federal Insecticide Fungicide and Rodenticide Act are based on the LD/50: "DANGER-POISON" means that 50% of the test animals were killed by a dose of 50mg. or less per kg. of body weight; "WARNING," by a dose of 51 to

500mg./kg.; "CAUTION" by 501 to 5000mg./kg.[1] However, even when the test is not mandated by government regulations, it is frequently used by manufacturers of drugs and consumer products on the supposition that it is a protection against proceedings for negligence.

Variables Impair the Accuracy
of LD/50 Tests

Recently, the LD/50 has been under increasing fire from scientists as well as humanitarians. At first glance, the test seems clear-cut since its end-point is an event as definite as death. But, as I have reiterated in this book, tests on animals are beset by numerous variables. Thus, while the actual death is definite, the moment it occurs is on a sliding scale: some die soon, some later. Furthermore, there are differences in susceptibility between species: penicillin is highly toxic in guinea pigs but generally nontoxic in humans. Some drugs, when tested by the LD/50, appear to be relatively nontoxic, even when given in large doses, but when administered in small doses over a longer period of time, prove to be killers. The value of the LD/50 may also be influenced by the age of the animal, its sex, state of nutrition and bedding material, also by the temperature and humidity, the time of the day or year the test substance is administered and by the method of administration. J.K. Morrison and co-authors, writing in *Modern Trends in Toxicology*, state:

"It cannot be denied, however, that a reproducible LD/50 determination can be achieved provided that experiments are

[1]
Comment by W.N. Scott (Universities Federation for Animal Welfare): "My chief objection is to people who continue to stuff animals with inert drugs until the substance almost comes out of their ears. Assessing toxicities of the order 5000mg./kg. is not killing the animal with the toxicity of the drug, but by distension of the stomach or something like that." (CRAE, 1977,p.30).

carried out under strigently controlled conditions guaranteeing
uniformity over all these numerous factors which contribute to
variations. But we are not convinced that such effort and experience are either economically or scientifically justified,
since the resultant information contributes little to the design
of future toxicity trials in animals, or to the assessment of
safety in man." (Morrison,J.,1968,p.12).

Stuffed to Death

There is no justification for using the LD/50 to test relatively nontoxic substances such as cosmetics and food additives.
The forced feeding of huge quantities of these may cause the
stomach to rupture or may kill the animal through the physical
insult to its system: this bears no relevance to the actual toxicity of the substance. It is claimed that this grotesque practice is being superseded by the "limit test", in which "a dose
of the substance at one percent of the animal's body weight
(equivalent to the ingestion of 1/4 lb. of face cream by a one-
year old child) is administered to a group of animals. If this
produces little or no toxic reaction, then the substance is pre-
sumed to be safe and no further testing is required." (Rowan,A.,
1979b,p.5). R.P. Giovacchini, whose tests at Gillette Laboratories are cited by the Food and Drug Administration as setting
standards for "adequately substantiating" cosmetics, is another
who cautions against LD/50s if there is a possibility of death
resulting from shock caused by "the sheer volume of material
ingested." To avoid this, he also recommends one percent of
body weight as the maximum dose. (Giovacchini,R.,1972,p.364).

Modifications of LD/50 Test Suggested
by Experimenters

Considering that the standard LD/50 may use from 60 to 100
animals, it is important to press for at least a reduction of
this wastage pending the phasing out of the test itself. A
suggestive comment comes from P.S. Rogers, Managing Director of

Hazleton Laboratories (Europe), one of the largest contract laboratories in the world:

"The test itself is usually run in two parts, involving a dose ranging test with two animals per group, followed by a repeat experiment using larger groups of animals over a narrower dose range....The LD/50 is a crude measure of toxicity. There is really little scientific justification for the test because reproducibility is not good, it can even vary from day to day, and the results are dependent on the animal strain used. *Sufficient information can usually be obtained from the dose ranging sector of the test.*" [Emphasis added]. (CRAE,1977,p.18).

Battelle's Pacific Northwest Laboratories have made some attempt to moderate the rigors of an LD/50-type experiment on ten beagles who received oral doses of a radionuclide which caused gastrointestinal lesions, diarrhea and death.

"To minimize pain the animals were sacrificed when clinical signs indicated death was imminent. To minimize the number of animals required for the experiments, the dose groups were completed in sequence so that data from one dose group could be used to plan sequential dose groups." (USDA/APHIS,1976,Battelle Memorial Inst.).

. . . . .

The above discussion was largely concerned with acute toxicity testing.

This section concludes with several paragraphs on more prolonged - subacute and chronic - test procedures.

Subacute Toxicity

In general this defines the biological activity of a compound, an estimate of the "no-effect" dosage, and the maximum tolerated dosage. The experiment usually covers a three month period. The animals generally used are, first, the albino rat, ten of each sex in each of six groups. One group receives no dosage (controls), the other five receive gradually decreasing dosage - starting from the maximum tolerated dosage which has been previously determined by a "range-finding" test.

Concurrently, dogs are used, of known species (usually beagles) and colony-raised, two males and two females on each of three dosage levels, plus a control group. The route of administration of the tested substances is as close to the proposed human usage as possible. Frequently in feeding experiments the animals will refuse the test material and then the stomach-tube is necessary. Biochemical tests are carried out both before and during the experiment; at the conclusion, the animals are killed and their organs and tissues studied. (Assn. of Food and Drug Officials, 1979,p.26).

Chronic Toxicity

Chronic toxicity tests are used to determine if a small dosage, apparently harmless on subacute testing, is toxic on long exposure. The Food and Drug Administration generally runs these chronic tests for two years. Over this length of time, larger groups of rats and dogs than used in the subacute tests are given graduated dosage so as to determine a "no-effect" level and a toxic level of dosage. At the end the animals are killed and autopsied. Sometimes much larger numbers of animals are used to be sure that there is no toxic effect. The FDA cites an instance where the use of 24 animals showed no effect at a certain dosage, whereas testing 200 rats at that dosage revealed an increase in liver weight among enough of them to be significant. (*Ibid.*,p.36).

3. COMPARISON OF TESTS IN INTACT ANIMALS AND CELL CULTURES

With chemical entities numbered in the millions and hundreds of new ones entering the home, the farm and the environment every year, the attempt to test all these toxicologically becomes overwhelming. The effect on humans is what we

need to know, but animals have been the traditional stand-ins.
However, as P.C. Rofe points out in an article reviewing the
growing use of tissue and cell culture in toxicology,

> "Toxic manifestations in the whole animal, be they changes
> in metabolic patterns or alterations in functional efficiency of
> specific organs, are still secondary to changes occurring within
> the cell...the facility of being able to observe the reaction
> of tissues from man under the same conditions as those of ex-
> perimental animals is a unique advantage."

Many systems using either animal or human cell or organ
culture, as well as plant material and microorganisms, have been
created with emphasis on those which are rapid, inexpensive and
can discriminate between those chemicals whose properties represent a high toxicity risk and those which are relatively innocuous. (Rofe,P.,1971).

In Vivo vs. *In Vitro*: Comparing Tests on Animals and Animal Cell Cultures

While there is no difficulty in proving that many chemicals
are toxic to cells in culture, it is more difficult yet essential
to demonstrate to what degree the cellular response is related to
the whole animal's response to the toxic compounds. Only if the
effect on the cells in culture can be shown to correlate closely
with the animal reaction - indeed can predict the toxic effects
in animals of other compounds of similar structure - are *in vitro*
tests on cells likely to supersede the *in vivo* ones in animals.
Or, if not supersede, then at least screen out the substances
which are acutely toxic and thus spare animals some of the LD/50-
type tests described on p.203-208.

As an effort in this direction, the work of Dr. Robert T.
Christian at the University of Cincinnati is interesting. According to a report in the *Christian Science Monitor* of Mar. 10,1978,
he has tested 30 chemicals on mammalian cell cultures and compared
his findings for them with tests of the same chemicals in the who

animal.   Toxicologically, Christian found that the results in the
two methods of testing agreed very well. (Salisbury,D.,1978).

Of course, if questions are asked about the transport of a
toxic compound in the body, about its metabolism or transformation into less or more toxic forms, its chronicity, etc., a cell
culture test which merely records the toxicity of the dose in
terms of inhibition of cell growth will be of little use.   One
must ask appropriate questions to get information of value from
tissue culture.

Autian and Dillingham asked the following question: given
a series of methyl- and halogen-substituted alcohols, each of
which is known to kill by a certain dose 50% of a group of mice
(LD/50) in seven days, can a similar effect be produced in cell
culture: i.e. 50% inhibition of cell growth (ID/50), with high
correlation between the two procedures?

They used a mouse fibroblast L cell culture. When the
ID/50s were calculated for the alcohols, giving the intrinsic
toxicity (T_i) of each, it was found that the correlation between
the T_i values and the LD/50s - between *in vitro* and *in vivo* -
was poor: there was a 164-fold variation. To reduce the gap, it
was necessary to introduce another factor into the equation: the
degree to which each of the alcohols separated either into the
aqueous or into the lipid (oily) compartment of the cell culture.
Expressed in terms of "lipophilicity," the most lipophilic alcohol was 60 times more "oil-loving" than the least. When the
equation was corrected for this, the 164-fold variation above was
reduced to 4-fold, which, allowing for variations due to statistical error, biotransformation and metabolism of the compounds,
is considered excellent. (Dillingham,E.,1973).

As the investigators point out, their alcohol study, although it provides meaningful information with respect to

structure-activity relationships and supports the basic equiva-
lence of *in vitro* and animal response,

"...only suggests the direction for the development of
good predictive systems. Without question...animal toxicity
parameters other than the LD/50 will be needed to deal with
different classes of biological responses. Tissue culture
parameters other than growth inhibition will undoubtedly be
needed."

They add that, as the list of chemicals tested in this
manner grows, mathematicians will have data from which to
predict the toxicity of new compounds and to determine what component parts of the molecule are contributing to the overall
toxicity. This is similar to the role of mathematicians described on p. 247.

In Vitro vs. In Vitro: Comparing Tests on Human and Animal Cell Cultures

An excellent review of toxicity testing in cell cultures
by Professor Roland M. Nardone of Catholic University of America
describes many of the advantages - and limitations - of the *in
vitro* method. But Nardone has come critical comments.

"The development of *in vitro* testing to date has been, at
worst, almost happenstance, and, at best, without taking full
advantage of the technical and theoretical advances of recent
years. Such amorphous growth is wasteful, inexcusable, and has
led to the publication of much research of questionable value."

To correct this, he suggests rigorous standardization of
test procedures, including the biological material used - cells,
tissues and organs, media and culture vessels.

"Serious consideration," he continues, "should be given to
the establishment of the toxic response of one or more well-
characterized 'reference' cell lines to diverse, ultrapure
reference chemicals drawn from several classes such as metals,
acetone, and National Cancer Institute standard carcinogens. The
response of such a test system could then serve as a yardstick or
standard against which other cells can be compared, against which
other toxicants can be compared, and for the continued assessment
of the reproducibility of a particular protocol." (Nardone,R.,
1977).

To study the subject of interspecies variations, P.C. Rofe has proposed that a range of adult human tissues be exposed to the action of selected compounds. "These compounds would be chosen for their contrasting toxicological effects and for the amount of data available about them. A selection of tissues from the principal laboratory animals would also be exposed under the same conditions." At the conclusion of the exposure, a sample of each tissue would be rapidly frozen and subjected to chemical and microscopic analysis, with emphasis on obtaining metabolic data. Since the liver is the main site of metabolism of foreign compounds, its study in culture would be of particular importance, and liver metabolites formed after exposing it to the test compounds could be added to the culture fluid of the other tissues. "From these studies it would be possible to design a routine procedure, the results of which would be available in the early stages of a major investigation, and would indicate not only the most appropriate species to use but also the most vulnerable tissues." (Rofe,P.,1971,p.683).

In other words, these investigators wish to standardize the use of tissue and cell cultures in toxicology. They would compare the chemical reactions of human and animal cells - under the same conditions - to a variety of compounds and their metabolites. From this would emerge routine *in vitro* procedures which would be reproducible, and the standardization would hopefully reduce the variables responsible for inconclusive experimental results and much wastage of animals.

Thirteen

TESTING SPECIFIC CHEMICALS

1. BACTERIA TEST OF WATER TOXICITY

The pollution of water is an ever-increasing concern, and has led to the development of biological testing methods using fish, water fleas and other aquatic organisms. Fish, for instance, are immersed in different concentrations of the effluent to be tested, and the percentage surviving after 96 hours is recorded. The concentration in which 50% are killed gives the LC/50. Beckman Instruments has introduced an alternative, the Microtox Acute Water Toxicity Monitor, using a strain of luminescent bacteria as the bioassay organism. The light-producing metabolism of the bacteria is about six times more sensitive to toxicants than fish lethality, and the decrease in light emitted can be measured by the Microtox quantitatively in half an hour as against 96 hours for the fish test.

Since each testing population of fish is potentially unique, they are much more difficult to standardize than bacteria; thus tests using the microorganisms are more reproducible. Also, they are said to be two orders of magnitude less expensive than the fish test (once the approximately $10,000 outlay for the instruments and stock of reagents has been made), and the bacteria test is certainly more humane. (Beckman Instruments,1979).

2. TESTING THE TOXICITY OF FOOD DYES AND INSEC-
TICIDES ON UNICELLULAR ANIMALS

A group of Japanese researchers led by F. Sako, at Hokkaido University, have shown that Paramecium caudatum, a unicellular animalcule, can be used to assess the toxicity of insecticides and food dyes. Comparing the use of small rodents and protozoa, they state:

"Although the LD/50 values for small animals have been generally accepted as indicators of the toxicity of food addi-tivies, use of such animals seems both time consuming and un-economical, judging from the fact that there has been a rapid increase in the numbers of various chemical compounds, including food additives, that require testing. Chlorella, Saccharomyces cerevisiae var. and Tetrahymena have [also] been used as indica-tors for these purposes." (Sako,F.,1977).

Dr. Oscar Frank at the New Jersey Medical School, East Orange, has also been working with protozoa, especially Tetra-hymena, devising assays for vitamins and amino acids in food-stuffs. He has shown these tests to be faster and cheaper than rat tests, and more sensitive. The American Fund for Alter-natives to Animal Research (AFAAR) awarded him a grant in 1979 for this research. (Frank,O.,1980).

3. CELL CULTURE TESTING OF MEDICAL DEVICES

A standard method of testing plastic and other "biomateri-als" is by implantation either in a rabbit's muscle or skin. W. Guess and associates devised an alternative test using a cell culture of mouse fibroblasts. These were covered with agar and neutral red (dye) and the sample of test material was placed on top. In 24 hours samples which were toxic killed the sur-rounding cells: these had lost the dye and were now colorless. After testing thousands of biomaterials in succeeding years and comparing the tests with the rabbit implants, it turned out that the tissue culture was actually more sensitive than the implant.

The U.S. Pharmacopoeia is considering this test as an alternative for the rabbit muscle implant procedure. (Guess,W.,1965).

In 1973, Wilsnack and associates reported a similar test on biomaterials, but instead of mouse fibroblasts they used a human cell culture (WI-38 embryonic fibroblasts). The authors point out that the human cells are free from the contaminating leukemia viruses often found in mouse or chick embryo cell cultures. They also proved more sensitive than the rabbit implant test. (Wilsnack,R.,1973).

4. TEST FOR EYE IRRITANTS

The use of the rabbit's eye to test not only cosmetics but detergents, pesticides and any substance which might be hazardous to human eyes is routine. This is known as the Draize test, after the Food and Drug Administration doctor who standardized it 35 years ago. Since it is a test which must cause extreme suffering it ought to be closely scrutinized and a humane alternative should be developed.

The rabbit's cornea is particularly sensitive: it is only 0.36 mm. thick compared with 0.51 mm. for both man and monkey. Furthermore, the rabbit's eye is not supplied with effective tear glands; thus the irritating substance cannot be readily dissolved. (Coulston,F.,1969).

The Consumer Product Safety Commission test for eye irritants can be found in the U.S. Code of Federal Regulations. (US Laws,Stats.,1979,*CFR*,Title 16,II,1500.42). One eye of an albino rabbit is used, the other, untreated, is the control. The substance to be tested - liquid, flake, granule, powder - is "gently" placed in the eye, and the ocular reaction is recorded at 24, 48 and 72 hours. The positive reactions, increasing in severity, are: "Ulceration of the cornea, opacity of the cornea;

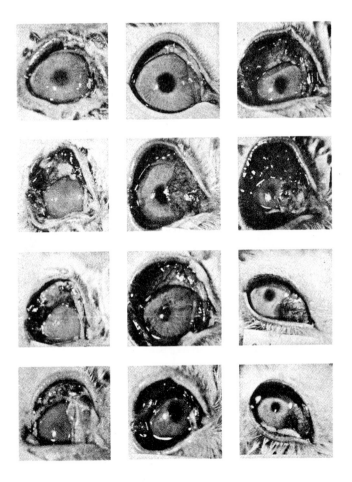

Fig.1. Chart showing reaction of rabbit's eye traumatized by
       test solutions.

inflammation of the iris, hemorrhage, gross destruction; or in
the conjunctivae an obvious swelling with partial eversion of
the lids or a diffuse crimson-red with individual vessels not
easily discernible." The accompanying chart (original in color)
depicts responses of varying intensity to specific test solu-
tions (Fig.1). Of course the rabbits are in restraint during

The familiar row of rabbits (USA)

Fig.2. Rabbits in an American laboratory restrained for Draize
       test. (Scottish Soc. for the Prevention of Vivisection).

the one to three days they are undergoing this suffering
(Fig.2).

        When animals are forced to undergo such suffering, the very
least one expects is that the tests will be reliable and the
results meaningful. A comprehensive study of these eye and skin
irritation tests, conducted in 25 cooperating laboratories (including Avon, Revlon and American Cyanamid-Lederle) by M. Weil
and R. Scala of Mellon Institute, Pittsburgh, and the Medical

Research Division of Esso, revealed "extreme variation" in the
way the laboratories evaluated the rabbits' reactions to standard irritants. The investigators concluded that "the rabbit
eye and skin procedure currently recommended by the Federal
agencies...should not be recommended as standard procedures in
any new regulations. Without careful reeducation these tests
result in unreliable results." (Weil,M.,1971).

More recently, investigators have attempted to recheck the
findings of Weil and Scala. F. Marzulli and D. Ruggles with the
cooperation of ten laboratories performed numerous Draize tests
on rabbits and confirmed that there were striking differences in
the way different technicians reported what they thought they
saw in the rabbits. Nevertheless, Marzulli and Ruggles insisted
that the test could still be valid for irritancy on the grounds
that only one of the four observed criteria had to be positive -
and the validity was still not discounted even if the observers
differed on which of the four that was! (Marzulli,F.,1973).

Another pair of investigators, although they claimed to
have found consistency in observations of irritancy, at the same
time noted large variability in the reactions of individual
rabbits. (Bayard,S.,1976).

Public Protests Against the Draize Test

The squirming in pharmaceutical circles to salvage an out-
worn, unreliable and inhumane test has not succeeded in silencing
the growing public criticism of this procedure.

A group of several hundred local animal welfare groups,
backed by the Humane Society of the United States and other
national groups, has been organized by Henry Spira. Its purpose
is apparent from its name, the Coalition to Stop Draize Rabbit
Blinding Tests. A campaign of public information and protest
has started, stressing the need for a humane alternative to the

Draize test, and directed at the Cosmetic, Toiletry and Fragrance Association, the federal agencies involved, and the leading cosmetic companies. These companies, starting with Revlon,
are being asked to allocate one-hundredth of one percent of
gross annual sales to finding such an alternative. (Anon.,1980a).

> *Court Ruling: Test on Rabbits' Eyes
> Inapplicable to Humans*

In 1974, the Northern District Court of Ohio handed down a
decision which has strengthened the public protests against the
Draize test. A girl using a shampoo suffered eye damage. Applying the Draize test, the Food and Drug Administration determined
that the shampoo was an irritant and brought suit against the
manufacturer. The ruling of the court, in favor of the manufacturer, was partly based on the finding that the FDA had failed
to show that the results of tests on rabbits' eyes can be extrapolated to humans. (*Ibid.*).

5. SKIN TESTS FOR TOXIC OR IRRITANT SUBSTANCES

The Consumer Product Safety Commission's recommended skin
tests call for acute (24 hours) or subacute (20 and 90 day) exposure. In the acute tests the animals, usually rabbits, have
the skin of the trunk clipped free of hair and abraded. They
are then anointed with the test substances, wrapped for 24 hours
in a close-fitting "sleeve" of rubber, and immobilized. In the
case of toxic materials, "dosage levels are adjusted...to be
fatal to 50% of the animals." (US Laws,Stats.,1979,*CFR*,Title 16,
II,1500.40-41).

Irritants are applied similarly, but in patches, and
evaluated according to the following scale after 24 to 72
hours.

Symptom	Description	Value
ERYTHEMA [redness]	None	0
	Very slight	1
	Well-defined	2
	Moderate to severe	3
	Severe (beet redness) to slight eschar formation (injuries in depth)	4
EDEMA [swelling]	Evaluation as above, from "none" to "severe" (4).	

Compounds which are only slightly toxic on acute exposure may be severely damaging on repeated application. Therefore, tests are recommended in which large doses of the substances are applied daily ("20-day subacute experiment") to areas of both intact and abraded skin. To abrade the skin, adhesive tape may be attached, then stripped off. Repeated applications of the tape allow the stripping of successive layers of skin down to the required level. (Lansdown,A.,1972).

Some experiments require the intact skin to be anointed daily for as long as 90 days. The skin is not covered with a rubber sleeve, as in the acute experiments, but the animals have to be in restraint and the potential suffering can be gauged by the comment: "the agent may produce a mummification or a coagulative effect on the skin...or complete necrosis, or severe eschar formation with desquamation." (Assn. of Food and Drug Officials, 1979).

"Pain and Distress": a Report from the Laboratories on Rabbits

Rabbits, as we have seen, are a favorite subject for testing irritants on eyes and skin, and cited below are excerpts from Annual Reports to the Dept. of Agriculture under the Animal Welfare

Act of companies which were among the largest users of these rabbits. The experiments described were all listed under Column D of the report form since they involved "pain or distress" for which pain-relieving drugs were not given.

Dow Chemical USA, Midland, Michigan. "1516 rabbits. The animals listed in the column headed 'Pain-No Drugs' were used in eye and skin irritation studies, such as those required by the Federal Hazardous Substances Act and by the Federal Insecticide, Fungicide and Rodenticide Act." (USDA/APHIS,1976,Dow Chemical Co.). *are required to be tested, but not! required to test on anim*

Procter and Gamble Co., Cincinnati, Ohio. "1561 rabbits. These are product safety tests required by the CPSC [Consumer Product Safety Commission] and FDA to evaluate the ocular irritancy and potential damaging effects of products after topical instillation in the cul-de-sac. It is necessary that no topical anesthetics or analgesics be used since their use may interfere in the proper evaluation of the test. Most products cause only transient irritancy and very mild damage to the conjunctiva and/or cornea. If severe irreparable damage is induced, the animal is humanely euthanized." (USDA/APHIS,1976,Procter & Gamble Co.).

Hill Top Research, Inc., Miamiville, Ohio. "1604 rabbits. The rabbits in our laboratory are often involved in several studies, sequentially. These include eye irritation studies, primary skin irritation and corrosivity studies, and finally a dermal LD/50 study. Generally, at least one of these studies involves a test substance that is painful to the rabbit. If a rabbit suffers ill effects as a result of a particular study, it is sacrificed and not used in any subsequent study." Pain relief omitted "because this would have interfered with protocol design and test results." (USDA/APHIS,1976,Hill Top Research, Inc.).

This last report gives a rare glimpse of repeated use of the same animals in painful experiments, which opens up a whole new dimension of suffering for these animals. How serious do the "ill effects" mentioned above have to be before the rabbit is put out of its misery?

"Professionally Acceptable"

By their own admission, these companies do not give pain-relieving drugs to the suffering animals. Why not? Because "it would interfere with the interpretation of the results." But the analgesic is to dull the sensation of pain; surely the results would still be plain to see on the blistering skin and in the hemorrhaging eye? No, the pain is declared "necessary," and the attending veterinarian, the laboratory director and the Animal Care Committee all certify that "professionally acceptable standards" have been maintained.

What these companies should be required to do is to produce scientific evidence that the use of analgesics *would* interfere with the results. If they cannot, then "unnecessary pain or distress" has been inflicted - and this is prohibited by Sec. 2.28(d) of USDA Regulations. (US Laws,Stats.,1979,*CFR*,Title 9,I).

.   .   .   .   .

To sum up: from the internal inconsistencies in these reports, and by comparing them with one another and with those of other companies, it appears that even the meager information requested under the Animal Welfare Act is not being supplied. These reports will continue to be useless until the regulations require that the number of animals used, *the nature of each experiment*, and the use or non-use of pain-relieving drugs (including postoperative analgesics) be specified.

Alternatives to Animal Tests for Eye Irritants

In 1975-76 Dr. P.B. Harper at Hazleton Laboratories Europe Ltd. investigated an alternative to the testing of irritants on the eyes of living rabbits. (Hazleton Labs. Europe Ltd.,1976 and 1977). The study was commissioned by the Dr. Hadwen Trust for Humane Research and involved the exposure of four different cell cultures (KB, HeLa, HEp2 and L929 cells) to shampoo-type products known respectively to be of low, intermediate and high irritancy.

These cell cultures were exposed to the shampoos for 72 hours, after which cells were counted to see how many had been destroyed and what had happened to the total protein content. It was found that the more irritating the shampoo, the greater the destruction of cells and the greater the decrease in total protein, the latter from interference with cell metabolism.

L929 (mouse fibroblast) cells showed great promise. These cells could distinguish between the different degrees of irritancy of all three shampoos. KB and HeLa, which are cultures of human cancer cells, were less sensitive, but could distinguish the low irritancy shampoo from those of intermediate and high irritancy (but could not distinguish between the latter two).

After this promising beginning, the investigation at Hazleton was discontinued, possibly because the Dr. Hadwen Trust could not continue financial support. However, an effort to develop an alternative test along these lines, without using animals, is currently being investigated at Johnson and Johnson pharmaceutical laboratories in New Jersey. Up to now, they have not been willing to divulge any details of their research, at least to this writer.

In addition to the above, a variety of techniques might be cited as alternatives: the use of human and animal corneas from eye banks, corneal tissue cultures, rabbit lenses prepared in organ cultures, and corneas, supplied by hospitals, which are frozen and may be kept indefinitely. (UAA,1973).

Alternatives to Animal Tests for Skin Irritants

When there has been exposure to toxic or irritating substances, the first question to ask is whether there is likely to be penetration of the skin and subsequent toxic effects in the body. If so, tests for general toxicity are necessary. If the substance is already being produced and is in use, these tests and epidemiological studies can be done on the workers in contact with the substance during the production and on those who are exposed to it after it has been distributed (e.g. farmers in the case of agrichemicals). (Gehring,P.,1973). However, if the substance, say as a pesticide spray, or detergent, or cosmetic, is in the phase of development, attention must first be focused on the skin itself and its relative impermeability or tolerance to the material.

Human volunteers are often used in testing cosmetics, detergents and the like, usually after screening on animals. However, animal tests can be misleading. Griffith and Buehler at the Procter and Gamble Company comment as follows on safety prediction for household products tested on humans, guinea pigs and rabbits: *why use animals*

"Though the guinea pig results were generally predictive of the human results, rabbit responses to the soap, the liquid detergent, and the cleaner would incorrectly place these products in the next higher, or moderate irritant category. Of the last five products, two (shampoo, pine oil cleaner) were weakly irritant, one (ammonia) was moderately irritating, and two (metasilicate detergent and bleach) were strongly irritating to human skin. Neither the guinea pig nor the rabbit skin would have predicted any of these responses correctly on the basis of absolute scores. Three of the rabbit results erred on the high side and two on the low side; the opposite was true for the guinea pig." (Griffith,J.,1977,p.158).

Griffith and Buehler used the "repeated-insult patch test" in their human subjects. A "scarification test," in which the test substance is placed on a small grid of scratches in the skin, under a $12mm^2$ chamber, is said to be faster and more accurate, espcially for mild topical agents. (Frosch,P.,1977).

Radioactive Isotopes Permit Systemic
Tests in Humans

It is true that the very distressing studies described on p.220-222 frequently involve the absorption of toxic products into the system over a comparatively long period; thus animals are used rather than man because of the potential danger. However, with proper safeguards, substances can be applied long enough in humans to allow systemic penetration and metabolic studies. R. Drotman tested a quaternary ammonium chloride widely used in laundry softener products, radiolabeled with ^{14}Carbon isotope, on the skin of several species - rat, rabbit, guinea pig and human - for 72 hours (144 hrs. for the human). He studied the systemic metabolic pathway of the product through the distribution of radioactivity in excreta (carbon dioxide, urine and feces). Since radiolabeled substances as small as one million millionth of a gram can be detected, the minute quantities of radioactive isotopes used were not dangerous. (Drotman,R.,1977, p.97).

When tissue damage is likely to be serious and put volunteers at risk, cultures can be exposed to the test material and cell injury can be observed microscopically. Time studies will show whether the trauma results in complete or partial cell destruction and, in the latter instances, with what degree of cell recovery. Non-specialized tissue may be adequate to assess the possibilities of cell survival, but if specific skin functions have to be studied, the human skin culture system used by B.A. Flaxman may be used. This culture includes proliferating epidermal cells and cells undergoing keratinization and desquamation. (Flaxman,B.,1972). Pieces of skin taken from human cadavers have also been used in permeability studies, but these are not always accurate predictors of percutaneous absorption in living man.

One reason is the absence of desquamation in the excised skin.
(Webster,R.,1977,p.120).

6. TESTING BY COSMETIC COMPANIES

Avon, as befits a large cosmetics company, uses many animals
in testing its products for "integumentary" (skin) irritation.
In addition to their 3,800 rabbits, they had 4,900 rats and 1,600
guinea pigs in 1973. (USDA/APHIS,1973,Avon Prod.) Along with all
approved laboratories in New York, they sign annually an agreement with the state in regard to the use of experimental animals,
agreeing, among other things, to their "humane handling" and that
"no greater number of animals will be used for scientific tests,
experiments, or investigations than is actually necessary." Describing work that would involve the use of living animals during
the year 1973, they stated:

"The work will involve the determination of acute toxicities
of cosmetic products or ingredients. Rats, guinea pigs, and rabbits will be used. Interest will be primarily in the oral and
percutaneous [through the skin] routes. In addition eye irritation and skin irriation tests as specified by the Food and Drug
Administration will be performed on appropriate species. Sensi-
tizing studies and inhalation toxicity studies will be done on
guinea pigs and/or rabbits. The emphasis in this research will
be on the integumentary system from a therapeutic, cosmetic or
irritation aspect." (Avon Prod.,1973).

Although from the description of the tests it is plain that
they must have caused much suffering, the company claimed in its
1973 Annual Report to the Dept. of Agriculture under the Animal
Welfare Act that none of these experiments "involved pain or
distress without the use of anesthetics, etc." In 1976, they
reported using 2,234 guinea pigs, 235 hamsters and 2,248 rabbits, claiming that these too suffered no pain. None of these
animals received pain-relieving drugs, but Avon admits that in
the same year another 346 rabbits did suffer pain or distress,

still without the use of anesthetics, analgesics or tranquilizers
("would interfere with test results"). The rabbits, they report,
were part of the "Safety Testing (Draize Eye Irritancy Test) of
all cosmetic products required by FDA." (USDA/APHIS,1976,Avon
Prod.).

Revlon, another large New York-based cosmetics company, con-
ducts similar tests, using 1,500 rabbits, 500 guinea pigs, 500
mice and 7,000 rats in 1972. A June 12, 1972 article in the
Christian Science Monitor, poetically captioned "White Rabbits -
Blue Eyes," described a visit to the five-story building containing Revlon's thirty laboratories and $2 million worth of
equipment. The reporter wrote:

"Some experiences tattoo themselves upon the memory. Three
white rabbits all in a row, encased in little boxes that look
like reducing machines. Helpless, with their heads sticking out
of holes in the top. Around the left eye of each what looks like
blue eye shadow." (Loercher,D.,1972).

In 1972 Revlon stated that 150 experiments were performed
on rabbits - "eye and skin irritancy tests on cosmetics" - and
140 on rats - "acute and subacute oral toxicity tests," and
these involved "pain or distress without anesthetics, etc."[1]
(USDA/APHIS,1972,Revlon Res.Center). The number of animals *per
experiment* was not given but could be quite large. In 1973
they again list numerous rabbits, guinea pigs and rats, and
state that no anesthetic, etc., was used in performing the
following tests: "Draize Eye Irritation, Acute Oral Toxicity,
Primary Skin Irritation, Subacute Dermal Toxicity and Inhalation
Toxicity," but now claim that none of these tests caused pain or
distress. Yet some of them are the same ones which they reported
as painful in the previous year. (USDA/APHIS,1973,Revlon Res.
Center).

1
 For the nature of some cosmetic oral toxicity tests, see p.207.

Finally, in 1976, Revlon reported a "total" of 1,600 guinea pigs and 2,000 rabbits used, but instead of a breakdown specifying, as the Dept. of Agriculture form requires, the number of these in the categories "no pain," "pain and (pain-relieving) drugs," or "pain - no drugs," they simply wrote under the "no pain" *and* "pain - no drugs" headings, "See Reverse." On the reverse is written:

"The following tests are run at this facility using the animals listed: a. Draize Eye Irritation; b. Acute Oral Toxicity; c. Primary Skin Irritation; d. Subacute Dermal Toxicity; e. Inhalation Toxicity; f. Guinea Pig Immersion Studies....No anesthesia is administered in any of the above procedures. This is due to the nature of the studies involved." (USDA/APHIS,1976,Revlon Res. Center).

Product Safety Has to be
"Adequately Substantiated"

The Food and Drug Administration denies that it "requires" testing of cosmetics (with the exception of colors in cosmetics, which have to be safety tested). However, it does require that the companies ensure that the safety of their products be adequately substantiated prior to marketing, otherwise they must be labeled: "Warning - the safety of this product has not been determined." (US Laws,Stats.,1978,*CFR*,Title 21,I,740.10).

The FDA has stated that "adequate substantiation" can be achieved by carrying out the testing procedures described in a 1969 article by R. Giovacchini of Gillette Research Institute. (Giovacchini,R.,1969,p.13).

Cosmetics - the Alternatives

In addition to the animals sacrificed or severely distressed in the tests above, there are others who suffer just as much in the production of cosmetics and soaps. A fixative is used to

prolong the fragrance of perfume, and the most popular is musk -
the secretion from a gland near the genital organ of civet cats,
beavers, Himalayan deer and rats. On a civet farm in Ethiopia,
the 30 pound animals are kept for years in wooden cages, 10 to
18 inches wide, 11 to 18 inches high, 3 to 4 feet long, from
which they are never released for exercise. It is denied that
scraping the anal glands every ten days to collect the musk is
painful, but confinement for seven to eight years in cramped
quarters, in a dark, smoky building heated up to increase musk
production, cannot be considered humane. Another fixative is
ambergris - it is produced by sperm whales and is either found
floating on the sea or is obtained from the intestines after
death. But this is minor compared to the great quantities of
whale oil required for soap, cosmetics, lubricants and margarine.
For this the whales are mercilessly hunted down. They are shot
by a harpoon carrying an explosive charge. Three seconds after
they are hit the charge is detonated inside their body. They
die agonizingly from massive internal injuries and hemorrhage.

These torments inflicted on the civet, the whale, the rabbit and the guinea pig are totally unnecessary. There are
satisfactory alternatives.

Both soaps and every type of cosmetic can be made entirely
from flower, herb and plant extracts. The few chemicals which
have to be used in lipstick and eye makeup can be nontoxic syn-
thetics. An organization in England, Beauty Without Cruelty,
with an American affiliate, has produced all of these. Since
the ingredients are harmless, it has been possible to test them
on humans exclusively, and safely. and the Food and Drug Administration has accepted them for importation into the U.S. There
are also a few small American cosmetic manufacturers who make

products, which, either in production or testing, have not involved animals.[1]

Reducing the Number of Tests - and Animals

Since the methods of cosmetic manufacture advocated by
Beauty Without Cruelty are not likely to be adopted overnight,
it is important to encourage other efforts too, providing they
lead at least to the reduction of animal testing. Andrew Rowan
describes such an attempt:

"The Cosmetic, Toiletry and Fragrance Association is trying
to reduce toxicity testing in the industry via their Cosmetic
Ingredient Review program. This program involves collecting research of various substances from the files of cooperating companies, and making these results available to companies planning
to use the same substance in a new product. In this way duplication of testing can be avoided and, consequently, many animals
can be spared." (Rowan,A.,1979b,p.6).

A panel of experts has commenced its review of the safety of
each ingredient in cosmetics (some 2,800 ingredients are known,
of which 119 were under review by Dec. 1979). The experts will
determine whether 1. the ingredient is safe as currently used,
or 2. that the ingredient is unsafe, or 3. that there is in-
sufficient information for the panel to make a determination.
The tentative report is then made publicly available for a 90-
day comment period. (Elder,R.,1979). Of course, a decision of
the third kind, above, may lead to renewed animal testing, but
decisions of the first and second kind, if accepted by regulatory
agencies, should reduce pressure on test animals.

A Brighter Future for the Whales?

Aid for the whales may be coming from another quarter.
Cetus Corporation, a biotechnology company in California, is now

[1] A current list of these can be obtained from Beauty Without
Cruelty, 175 West 12th Street, New York, N.Y. 10011.

"working on modifying oil with microorganisms to make certain fine lubricants which could replace oil from the threatened sperm whale." (Wilkinson,J.,1980). And Witco, a British company, has produced a synthetic material, called Cyclol SPS, based on cetyl palmitate and other vegetable sources, which has the same properties as spermaceti, the main substance derived from whale oil. (Spermaceti is used to give cream cosmetics a firm appearance and to make them flow easily). (Anon.,1980h).

7. TESTING ANALGESICS

The standard tests to determine the value of potential pain-relieving drugs are skin-twitch and tail-flick (rats and mice) using radiant heat, the hot-plate, and the "writhing test." In the last-named, "mice are given phenylquinone, intraperitoneally, and after a few minutes begin to writhe and stretch their abdomen from injection of this irritant." (USDA/APHIS,1973,Warner-Lambert Res.Inst.). Since the movements are eliminated by low doses of narcotics, and by other pain-relieving drugs, it is a test of analgesia.

J. Cochin, reviewing methods for assessing analgesia, says: "The 'writhing test,' which is also euphemistically called the 'mouse peritoneal test,' has caused a great deal of furor, since writhing seems to offend the sensibilities of many investigators." (Cochin,J.,1973,p.704). Endo, in Garden City, New York, states: "minimal discomfort is experienced by unanesthetized mice and rats during exposure to phenylquinone writhing." (USDA/APHIS, 1976,Endo Labs.). Although the discomfort may be minimal in the opinion of the Endo spokesman, the Animal Welfare Institute comments: "This particular test has been under fire for many years....The Universities Federation for Animal Welfare urged its abolition some ten years ago." (AWI,1973,p.2).

Pinching the skin, the end-point being a squeal, is also used, and an "ingenious variation" is described by J. Cochin (Cochin,J.,1973), whereby the sensitivity of the rat to the pressure stimulus is increased "by injecting a yeast suspension into the hind paw and applying pressure to the inflamed area produced by the yeast injection...a struggle reaction being used as the end-point rather than vocalization." He also mentions "electrical methods, such as shock to the tooth pulp, or to the scrotal sac, or to the tail," but these give inconsistent results. More complicated methods have been devised to make the animal feel not only pain but fear - "conditioned anxiety, a concept that must raise some eyebrows of investigators who find it difficult to conceive of anxious rats."

This last comment is not a joke. Reflecting the attitude of this type of investigator, the Dept. of Agriculture actually issued a regulation stating that animals were not known to feel anxiety: "The word 'anxiety' is a psychiatric term that is only applicable to humans." (US Laws,Stats.,1979,*CFR*,Title 9,I).

Tooth Pulp Stimulation

"Seventy-one rabbits had anesthetics for implantation of electrodes into tooth pulp of upper incisor teeth. Tooth pulp was subsequently stimulated electrically to test compounds for analgesic effects with no anesthesia." (USDA/APHIS,1976,ICI-US). This statement appeared in the 1976 Annual Report to the Dept. of Agriculture of Imperial Chemical Industries. The nature of this implantation is described in an article in the journal *Pain*:

"Electrical stimulation of intradental nerves presents no special problem if, as in animal experiments, a cavity can be cut through the enamel and dentine to allow electrodes to be applied directly to the inner dentine or to the pulp surface, where the nerve endings are situated. In human experiments this is not usually possible and stimuli have to be applied through the enamel and outer dentine." (Matthews,B.,1976).

"No special problem" to the experimenter, but to the animals...?

During an Aug. 1978 visit to Huntingdon Research Center, Huntingdon, England, a subsidiary of Becton Dickinson, I saw a group of rhesus monkeys who had received tranquilizers and whose perception of pain was being tested by tooth pulp stimulation. Wires were led from a tooth, subcutaneously and through the cheek, to emerge just back of the ear, where they were connected with electrodes. The monkeys were all restrained in wooden collars and body stocks. The nearest one kept looking at me apprehensively and opening his mouth; every time the tooth was stimulated he twitched his lips. "It's not really painful," said the technician in a reassuring tone. Even if the sensations of the monkey as he was being shown to the visitor were at that moment only mildly distressing, what may be only mildly distressing may become very disturbing if indefinitely repeated. I failed to inquire whether the monkey had anything to look forward to besides lifelong restraint and a perpetual toothache.[1]

Tooth shock has become a fashionable experimental procedure only since the 1960's. In an article by K.A. Anderson and others, the following comment appears:

"The nature of the pain that resulted from electrical stimulation of human teeth was reported to be excruciating, poorly localizable pain that generally persisted for some time after the pulpal stimulus was terminated and can, therefore, be classified as diffuse pain." (Anderson,K.,1976).

It was concluded that in monkeys and cats, close to man on the phylogenetic scale, the effects would be similar.

[1]
I was unaware that during the year of my visit to Huntingdon, the laboratory had killed 1342 primates in acute toxicity experiments. The Int'l. Primate Protection League has drawn attention to this, and to the painful death there of 10 crab-eating macaques, poisoned in an experiment with the weed-killer diquat. (Anon.,1980c).

Alternatives in Analgesic Testing

The relatively harmless character of analgesics - a fact which could be confirmed by safety tests on human cell cultures - should make them suitable for testing on human subjects. Barring veterinary use, this would also prove a shortcut to the ultimate beneficiary, and would eliminate confusion from interspecies extrapolation.

Cochin points out that although the "writhing test" is sensitive to the analgesic action of the narcotic antagonist drugs, one of its

"major drawbacks is its lack of specificity, since many nonanalgesic drugs also inhibit writhing. Also there appears to be a lack of correlation with potency of analgesics in man. It is of interest to point out that the analgesic activity of the narcotic antagonists is one property of this class of compounds that is relatively easy to demonstrate in man (Eckhardt, E.,1958) and extremely difficult to show in almost any animal model that can be devised." (Cochin,J.,1973,p.704).

It is only fair to add that "human guinea pigs" have individual variations in sensitivity to pain, which can be affected by suggestibility, the environment and the motivation (cf.p.14). If animals have to be used, perhaps the hot-plate test is relatively benign. In this, a rat or mouse which has been injected with a drug to be tested for analgesic potency is placed on a plate which is heated to a constant temperature: 45° - $55^{\circ}C.$, for instance. The time to the end-point is a lifting of the hind leg or licking the front paws. If the animal is densensitized to pain by the analgesic, there is an automatic switch-off of the heat after so many seconds to prevent tissue damage. Since tissue damage would confuse the results - the animals are used repeatedly in what are usually long-term studies - the animals' feet must be protected.

However, there is also the possibility of using organ cultures, especially ones of the guinea pig ileum and the mouse

vas deferens, without the disadvantage of involving the whole
animal from which the drug, once administered, cannot be readily
removed. Kosterlitz and associates found that the "correlation
between the values obtained by the guinea pig ileum assay and
those of the mouse hot-plate test was always very close."
(Kosterlitz,H.,1975).

8. NEW DRUGS: DEVELOPMENT, AND TESTING FOR
SAFETY AND POTENCY

Medicines are meant to alleviate or cure pathological conditions. To induce such conditions in test animals which are
healthy to begin with it is necessary to make them ill by such
procedures as the injection of pathogenic organisms or of cancer
cells; the application of irritants to eyes, skin or mucous mem-
branes; the inhalation of substances producing choking or bronchospasm; stressing by electric shock or other means to produce
fear, depression and gastrointestinal ulceration; inducing pain;
causing coronary occlusion and other cardiovascular pathology;
producing burns; creating nutritional deficiencies, and performing other types of experiments all of which cause considerable
distress or pain which may or may not be alleviated by the drugs
being developed.

When procedures have caused pain in certain "reportable" animals, the Animal Welfare Act requires that these be specified in
the research facility's Annual Report to the Dept. of Agriculture.
If the pain or distress was not relieved by an anesthetic, analgesic or tranquilizer, the facility must explain why the appropriate drug was not given.

Experiments at Warner-Lambert Research Institute

Reportable animals are dogs, cats, hamsters, guinea pigs,
monkeys, rabbits, "wild animals" and marine mammals. Since rats

and mice need not be reported, and since the vast majority of
experiments are performed on them, the reports are very limited
in scope. However, for some unknown reason, the Warner-Lambert
Research Institute, which is connected with the large pharmaceutical company in Morris Plains, New Jersey, included rats and
mice in their Annual Report for 1973. They also gave unusually
explicit details of the experiments and the pain caused, and
explained why pain-relieving drugs were not used. These clear
descriptions of experiments typical of those performed in all
large pharmaceutical laboratories are exactly the way these organizations should report procedures associated with unrelieved
pain, but judging by my inspection of hundreds of these Annual
Reports forwarded to the Dept. of Agriculture, the agency practically never gets anything as straightforward as this, especially with the gratuitious but most informative mention of rats and
mice. (USDA/APHIS,1973,Warner-Lambert Res. Inst.).

"Warner-Lambert Research Institute
170 Tabor Road, Morris Plains, N.J. 07950

Annual Report for 1973 to USDA/APHIS of Animals Used in
Research or Experimentation.

B. Names of Animals	C. Approx. No. Used	D. No. of Experiments* Involving Pain or Distress without Use of appropriate Anesthetic, etc.
Mice	120,000	181
Rats	65,000	275
Hamsters	5,000	4
Guinea Pigs	27,000	188
Rabbits	1,000	12
Cats	300	
Dogs	2,000	197
Monkeys	15	

*Unlike the new form issued in 1974 - cf.p.240 - this form
does not elicit number of animals used per experiment.

"E. Reason for not using appropriate Anesthetic, etc.:

1.   Experiments which are performed by the Toxicology Dept.
involve a search for untoward reactions at high dose levels.
The concurrent use of anesthetics, etc. would mask the reactions we are attempting to note. In general the number of
animals showing distress from such studies amount to relatively few as compared to the numbers of animals used.

2.   Approximately 100 experiments using 5,000 mice are conducted yearly in evaluating the analgesic properties of
potential agents. Mice are given phenylquinone, intraperitoneally, and after a few minutes begin to writhe and
stretch their abdomen from injection of this irritant. Other
techniques required to induce pain in rodents (i.e. hot plate,
tail flick) are used as secondary procedures in analgesic development.

3.   Approximately 50 experiments are done yearly using mice as
objects for mouse-killing rats. This test is quite sensitive
for anti-depressant agents and is considered one of the more
reliable and predictive animal models. Usually mice previously used for some other experiment are employed.

4.   Mice are used in approximately 25 experiments a year in
which aggressive behavior is elicited by means of isolation
or painful electric foot shock. Drugs are studied in attempts
to prevent this display. Such procedures are important in the
development of tranquilizing agents.

5.   Two hundred experiments a year, comprising thousands of
rats, are conducted in conscious subjects given a phlogistic
agent into a rear paw. This treatment induces pain and swelling at the site of injection (carrageenin) or at a distal site
as well (adjuvant arthritis). Such experiments are essential
in the screening and development of potential anti-inflammatory
agents.

6.   Rats are used in approximately 25 experiments a year in
which aggressive behavior is elicited by means of isolation
or painful electric foot shock. Drugs are studied in attempts
to prevent this display. Such procedures are important in the
development of tranquilizing agents.

7.   Approximately 10 experiments a year using a total of 200
rats are conducted in evaluating the anti-ulcer properties of
potential agents in a Stress Ulcer Model. Rats are restrained
in wire mesh and suspended in a water bath (chest deep) for a
period of time. Vocalization and gastric hemorrhage occur
and drugs are studied in attempts to prevent these gastric

lesions. Since stress is thought to be an important mechanism for causing human gastrointestinal ulceration, we attempt
to duplicate this etiology in animals.

8. Our Chemotherapy Section uses approximately 300 guinea
pigs annually in experiments involving the abrading of shaved
skin to produce dermatophytic infections. These procedures
are necessary in order to discover and evaluate potentially
useful agents for the treatment of fungal infections.

9. For bronchodilator screening approximately 1,800 guinea
pigs are subjected to an aerosol of histamine or methacholine.
If animals are not protected by a bronchodilator they develop
intense bronchospasm and will ultimately expire of asphyxia.
These studies must be done in the unanesthetized state in
order to determine the end-point.

10. We use approximately 180 dogs that previously have had one
coronary artery ligated under surgical anesthesia. They develop a ventricular arrhythmia and are used for screening
anti-arrhythmic agents. On some occasions the drug may produce an effect which causes distress, e.g. convulsions. Under
these circumstances the animal is usually given a barbiturate.

> Edward Schwartz, VMD, PhD, Director, Dept. of
>   Toxicology
> Henry H. Freedman, PhD, Director, Biological
>   Research
> Richard C. Brogle, PhD, Director, Clinical and
>   Regulatory Services"

The first paragraph above refers to studies of "untoward
reactions at high dose levels" of drugs. This is a kind of toxicity testing - essentially the overdosage approach - well illustrated in an experiment at the University of Rochester, New York,
by B. Weiss and others, reported under the title "Movement disorders induced in monkeys by chronic haloperidol treatment."
(Weiss,B.,1977). Haloperidol is a major antipsychotic drug,
marketed by McNeil Laboratories, and in common with many "neuro-
leptics" may produce movement disorders as side effects. In
this experiment the monkeys, after several months of treatment,
began to exhibit "violent, uncontrolled movements that flung the
animals about the cage." Some exhibited writhing and bizarre

postures. These symptoms, in a milder form, have been observed
in humans treated with antipsychotic drugs. In some cases they
become persistent and irreversible, even when the drugs are discontinued.

Paragraph 2 refers to the testing of analgesics. For
further discussion of this, cf.p.232 ff.

Prey-killing experiments (para. 3) have been described in
my section on "aggressive" behavior, p.32 ff., also aggression
resulting from painful electric shock or isolation (paras. 4 and
6) was discussed there on p.37 ff.

For gastric ulcers caused by stress (para. 7), see J.
Weiss's use of shock, p.64, and S. Boyd and associates' use
of shock and cold restraint, p.65.

The bronchodilator experiments of para. 9 were discussed in
the chapter "Inhalation of Toxic Substances," p.104 ff.

Changes in the Annual Report Form under the Animal Welfare Act

Since the year of Warner-Lambert's report, above, the Dept.
of Agriculture has introduced (1974) a revised Annual Report form
under the Animal Welfare Act. It requires, in addition to the
number of animals suffering pain or distress who received no
pain-relieving drugs, a list of the reportable animals suffering
pain or distress who did receive drugs, and a list of the animals who suffered no pain or distress. As before, the form asks
for an explanation of why pain-relieving drugs were omitted in
the case of animals suffering pain or distress. While the old
form merely asked that the number of experiments causing unrelieved pain be given, the new (1974) form asks for the actual
number of animals used.

Listing the total number of reportable animals used rather
than merely the number of experiments is an improvement on the

old form, but since the research facility is not required to
specify the nature of the experiments, it is still impossible
to evaluate statements about the presence or absence of pain
or distress and the consequent giving or withholding of pain-
relieving drugs. What, for example, is one to make of the A.H.
Robins Co., which like other large drug houses manufactures
antibiotics, antihistamines, analgesics and tranquilizers?
Robins reports the use (including rats and mice) of 34,026 animals in 1976, but gave no pain-relieving drugs, and states that
none of these animals suffered any pain or distress. (USDA/APHIS,
1976,Robins,(A.H.)Research Labs.). How did they do it? When
one compares their statements with the Warner-Lambert Research
Institute's 1976 Annual Report, *Table 1*, (USDA/APHIS,1976,War-
ner-Lambert Res. Inst.), one finds that the latter lists some
thousands of animals (this time omitting rats and mice) requiring pain-relieving drugs or suffering pain and distress (Column
D).

   Warner-Lambert's 1976 explanations of why pain relief was
withheld (keyed by index numbers to Column D) are much briefer
than they were in the 1973 report, but they are still explicit:

   "1. Experiments which are performed by the Toxicology Department involve a search for untoward reactions at high dose
levels. The concurrent use of anesthetics, etc. would mask the
reactions we are attempting to note. In general the number of
animals showing distress from such studies amount to relatively
few as compared to the numbers of animals used.

   2. Safety studies of drugs on gross behavior or on electrical activity of the brain, in which distress may have been
induced; required by FDA for potential side effects.

   3. Aggression produced by electrical stimulation of brain;
recognized technique for developing new antidepressants.

   4. Animals subjected to bronchoconstriction due to histamine spray for screening and development of new drugs.

   5. Avoidable and escapable shock schedule (Sidman) for
testing tranquilizers." (*Ibid.*).

ANIMAL AND PLANT HEALTH INSPECTION SERVICE
VETERINARY SERVICES

ANNUAL REPORT OF RESEARCH FACILITY
(Required For Each Facility (Site) Where Animals Are Held)

January 24, 1977

2. HEADQUARTERS RESEARCH FACILITY
(Name and Address, as registered with USDA, Include Zip Code)

Warner-Lambert Research Institute
170 Tabor Rd., Morris Plains, N.J.

INSTRUCTIONS: REPORTING RESEARCH FACILITY *(SITE)* complete items 1 through 26 and submit to your Headquarters Facility. Attach additional sheets if necessary. **HEADQUARTERS FACILITY** complete items 27 through 29 and submit on or before February 1 of each calendar year to the Area Veterinarian in Charge for the State where the research facility headquarters is registered.

3. REGISTRATION NUMBER

22-2

4. REPORTING FACILITY *(Name and Address, Include Zip Code)*

Warner-Lambert Research Institute
170 Tabor Rd., Morris Plains, N.J.

5. REPORT OF ANIMALS USED IN ACTUAL RESEARCH OR EXPERIMENTATION

Section 2.28 of the Animal Welfare Regulations requires the appropriate use of anesthetics, analgesics, and tranquilizing drugs during experimentation. Experiments involving necessary pain or distress without use of these drugs must be reported and a brief statement explaining the reasons.

ANIMALS COVERED BY ACT	NO PAIN Number of Animals Used Where No Pain, Distress, Or Use of Pain Relieving Drugs Was Involved.	PAIN AND DRUGS Number of Animals Involving Pain or Distress Where Appropriate Anesthetic, Analgesic, or Tranquilizer Was Used.	PAIN - NO DRUGS Number of Animals Involving Pain or Distress Without Use of Appropriate Anesthetic, Analgesic, or Tranquilizer. *(Attach brief explanation.)*	TOTAL
A	B	C	D	E
6. Dogs	130	1340	98 [1] [2]	1568
7. Cats		110	17 [3]	127
8. Guinea Pigs	12	1240	3328 [4]	4580
9. Hamsters	2046		328 [1]	2374
10. Rabbits	747	16	50 [5]	813
11. Primates	2 Rhesus 8 Squirrel	3	85	98
WILD ANIMALS *(Specify)*				
12.				
13.				
14.				
15.				
16.				
17.				

CERTIFICATION BY ATTENDING VETERINARIAN OF RESEARCH FACILITY OR INSTITUTIONAL COMMITTEE
I (We) hereby certify that the type and amount of anesthetic, analgesic, and tranquilizing drugs used on animals during actual research or experimentat deemed appropriate to relieve all unnecessary pain and distress for the subject animals.

18. SIGNATURE OF ATTENDING VETERINARIAN *Donald Abrutyn, V.M.D.*	19. TITLE Director, Department of Toxicology	20. DATE 1/24/77
21. SIGNATURE OF COMMITTEE MEMBER	22. TITLE	23. DATE
24. SIGNATURE OF COMMITTEE MEMBER *Elliot Steinberg*	25. TITLE Director, Research Administration	26. DATE

CERTIFICATION BY HEADQUARTERS RESEARCH FACILITY OFFICIAL
I certify that the above is true, correct, and complete and that professionally acceptable standards governing the care, treatment, and use of anesthetic appropriate use of anesthetic, analgesic, and tranquilizing drugs, during actual research or experimentation is being followed by the above research faci sites (7 U.S.C., Section 2143).

27. SIGNATURE *(Responsible Official)*	28. TITLE	29. DATE

VS FORM 18-23   REPLACES ANH FORM 18-23 (5/72) WHICH IS OBSOLETE

Table 1. Warner-Lambert's Report to USDA/APHIS for 1976

The Effect of Drugs on Behavior

The first paragraph, above, repeats that in the 1973 report, and paragraph 4 is a repeat of the bronchodilator drug testing. The other tests are similar to the 1973 ones using behavioral evaluation of drug effects for which references in this book were given on p.240. These references were mainly to basic studies in behavioral psychology in which an animal is first trained to behave in a certain manner - to act aggresively under the stimulus of painful electric shocks, or to develop a stereotyped pattern of food avoidance when subjected to a conflict situation in which food freely offered alternates with food accompanied by electric shock. Once a characteristic behavior pattern has emerged it can be used in a pharmacological investigation of a new drug. If the drug is a tranquilizer, the animal may become relatively indifferent to the shock, with a decrease in the food avoidance rate. If the drug is an antidepressive stimulant, the previously induced aggressivity may increase. This gives a rough approximation of how these drugs affect certain behaviors.

That there are disadvantages of drug testing based on behavior shaped by painful stimuli was pointed out on p.74 ff. The many variables inevitable in the behavioristic approach and the disturbing effect of fear and pain often introduce confusion into the results. Why are the tests done in this fashion, then? The answer may be in a phrase quoted in paragraph 2 of Warner-Lambert's 1976 report: "Required by Food and Drug Administration." Have both the experimenters and the regulatory bodies lost sight of the inevitable variability of animal subjects in their preoccupation with ingenious technology and impressive statistical "verification"?

There may be better ways of arriving at the desired information. The science of drug design, as described in the next chapter, predicts the effects of therapeutic substances with precision sufficient to greatly reduce, if not eliminate, much of this painful and often hit-or-miss work with animals.

Fourteen

THERAPIES OF TOMORROW

1. DRUG DESIGN

The number of new drugs and chemical products which appear every year on the world market is enormous. A variety of factors stimulate this flood: competition between the manufacturers; the feeling of the general public that for every ill there's a pill; the promotion through advertising of patent medicines, cleaning products, cosmetics, pesticides (there are 35,000 pesticide formulations alone).

When it comes to drugs a vast amount of careful testing is required both to protect the public and the pharmaceutical manufacturer. Dr. W.G. Richards describes the process roughly as follows: "A research group finds a compound which will produce a biological response of a useful sort. They then get chemists to synthesize similar compounds and these are tested on animals until a better potential drug is found." He calls it "little more than inspired hit or miss" and, since thousands of drugs are tested for the one that eventually reaches the market, "wasteful of time, money and a great cause of suffering." (Richards,W.,1975).

As an alternative to this, there is the application of what has become known as the science of "drug design," an attempt to correlate the observed biological response with the molecular structure of the drug. The goal is to "predict the

biological activity of the molecule prior to evaluation or even
synthesis in order to reduce the costly and time-consuming
synthetic work" - and to minimize screening with animals. Why,
asks Dr. Richards, kill 2,000 rats or monkeys trying to understand the action of a molecule or two in the whole effect of a
drug when mathematical calculations can tell us what are all the
possible effects in advance? Then, if animals are still needed
to confirm the results, no more than 50 may be required to produce the desired information. Since Dr. Richards, a chemistry
lecturer at Oxford, is a leader in theoretical pharmacology, he
can take us far enough into this difficult subject to open it up
for us, even though we shall have to part company before the
bristling mathematical equations appear.

"The conventional wisdom," says Richards, "is that drug
action at the molecular level involves the interaction of the
drug, a small molecule, with a macromolecular receptor." On the
latter is a binding site waiting for a drug with molecular structure and an electrical charge which will fit. The small molecule is flexible, however, changing its shape (or conformation
of its atomic nuclei) and the distribution of its electrons to
suit different receptors. Richards says: "A drug receptor will
experience the effects of the electron clouds of the approaching
small molecule," and theoretical calculations, supported by experimental studies such as nuclear magnetic resonance spectros-
copy, are capable of indicating the range of possible confor-
mations of a drug's active molecules which might be suitable for
binding. (Richards,W.,1977).

A Tailor-made Drug Against Mouse Cancer

The story of how the drug azetomicin, which is a cancer-cell
killer in mice, was developed, illustrates the above. As described in an Apr. 3, 1978, news item in *The New York Times*, the

scientists who synthesized the drug - Dr. Martin A. Apple and
associates - did so by looking at a range of drugs which used
the basic genetic material of cancer cells - DNA - as receptors.
They obtained basic information with the aid of PROPHET, a
nationwide computer which has a vast store of chemical and biological data. They then simulated molecules in three dimensions
on television screens and moved them around until they got con-
formations which they knew would adhere for comparatively long
periods to the receptors on the surface of the cancer cells.
While not yet safe to test on humans, azetomicin in mice has
proved to be one of the most potent anti-cancer drugs ever made.
(Anon.,1978c).

Computers in Drug Design

R. and M. Harrison describe the value of computers in drug
design. "Computers have been applied in pharmacology in several
areas. They have been used together with models of drug actions
in simulating and inhibiting physiological responses to predict
effects of individual drugs and combinations of drugs in various
doses. There is even a science of quantum pharmacology which
makes quantum chemical calculations and correlates factors such
as the strength and configuration of various chemical bonds within a drug molecule with the pharmacologic actions." (Harrison,R.,
1978).

New Compounds Designed to Fit a Receptor
Site in Hemoglobin

Another experiment in drug design relies on knowledge of
the structure of the receptor site in human hemoglobin. A small
molecule with a long name, 2, 3-diphosphoglycerate (DPG), inter-
acts with this receptor site and causes the hemoglobin in human
blood cells to lose some of the oxygen it has an affinity for.
C.R. Beddell and associates at Wellcome Research Laboratories in

England used a scale model of the receptor site and designed
three new compounds with atoms arranged to bind with those in
the site.   These compounds predictably promoted oxygen liber-
ation, like DPG.   This method was radically different from the
old-fashioned practice of starting with a substance which has
a biological property, say that of liberating oxygen, then
synthesizing similar compounds with this property and testing
them on animals to see which one works the best.   Beddell's
group, on the other hand, concentrated on compounds which fitted
the receptor site.   These in fact turned out not to be closely
related in structure to the natural starting substance DPG, even
though they shared its oxygen liberating property. (Beddell,C.,
1976),

Molecular Structure Visualized by
Electron Microscopy

One way of studying the molecular structure of receptors is
with the electron microscope, which can visualize objects of the
order of 5 - 10 angstroms.   This is within the range of molecular
dimensions.   Walton and Bucley observe in their 1975 paper:

"As interactions of cells with molecules having known structures are studied and understood, it will become increasingly
possible to predict on the basis of its molecular structure the
changes that a molecule or newly designed drug will induce in
cells.   This vast but largely unexplored area of toxicology
promises important developments which are basic to the understanding of cell toxicity and to the design of effective therapeutic agents." (Walton,J.,1975).

2. "GENE SPLICING"

Recombinant DNA techniques or "gene splicing" have made it
possible for scientists to move genes between species; for example, to take one of these biologic units of heredity from a
disease-causing virus - actually to cut it out with a specific

enzyme - and to insert it, again with the use of an enzyme (the microbiologist's scalpel) into a bacterial cell. By this process, the nature of the gene and the deoxyribonucleic acid (DNA) of which it is composed can be studied, both as to structure and function. Meanwhile the bacteria, like a little factory, will start replicating the foreign gene in abundance, and via RNA (the other nucleic acid molecule) the amino acid code of the gene can be "translated" and expressed in the form of protein. If the particular gene selected is the one that expresses the surface antigenic protein - the one that elicits a human immune respone in the form of antibodies - a vaccine against the disease has been synthesized.

If the long coils of DNA molecules are merely extracted from the animal cells and not reinserted in the bacterial cells, the DNA will break up into fragments and the genes cannot be isolated without contamination. But if they are introduced into certain components of the bacterial cell, which can be either plasmids (rings of extra-chromosomal DNA) or bacteriophage (viruses that live inside the bacteria), then the inserted material can be cloned without impurities, since plasmids and bacteriophage have very few genes compared with the whole bacterium which has several thousand. (Gilbert,W.,1980).

Synthesis of Hepatitis B Vaccine

Researchers at several European centers and at Stanford University in California have actually cloned a surface protein of the hepatitis B gene in E. coli bacteria, and the production of the antigen is now possible. (Anon.,1979b). This does not mean that a commercial vaccine is available - there is much testing and refining still to be done - but it is clear that this will be accomplished sooner or later. Such a vaccine would be far superior to one produced from whole virus (as in the

present polio virus technique) for the following reasons. 1. By
starting with genetic material from a human donor, no animals
with the variables and contaminating germs they invariably introduce would be involved in the production of the vaccine.   Thus
it would be scientifically superior.   2. No animals' lives would
have to be sacrificed in the production phase, thus it would be
humane.   3. When the whole virus, even if inactivated or attenuated, is present in the vaccine, there is always a potential of
virulence and the transmission of the disease.

As explained in the discussion of polio vaccine production
(cf.p.195 ff.), much of the testing of such products - testing
inevitably associated with suffering of test animals - is to
guard against the disease-producing elements of the virus.   But
these are absent when only the gene for the antigenic surface
protein has been extracted from the DNA.   Thus there would be
no need to test the final product, the antigenic vaccine, for
virulence, since it would lack the capacity to cause the disease although retaining the power to elicit immunity in the individual receiving it.   The potency of the vaccine can be tested
immunologically - *in vitro* - by radioimmunoassay techniques
against a radiolabeled standard vaccine of known potency.   The
unlabeled vaccine will compete with the radiolabeled standard for
binding sites on their antibody.   The lower the percentage of
radioactivity bound to the antibody (by precipitate count) the
higher the potency of the unlabeled vaccine.

One of the unpleasant complications of persistent infection
with hepatitis B is cancer of the liver (hepatoma).   This, however, has led to an interesting discovery.   Arie Zuckerman at
the London School of Hygiene and Tropical Medicine has found
that a viral-type protein on the surface of liver cancer cells
in people who are also carriers of hepatitis B is very similar
to the viral protein of the hepatitis antigen in their blood.

Both of these antigens will stimulate resistance to the disease, but cell cultures can be made from the cancer, stored in a deep freeze, and there they will go on producing the viral antigens indefinitely. From these, carefully extracted to avoid any risk of contamination from the cancer, large-scale vaccine production may get under way even before the gene-splicing method is generally available. (Anon.,1980b).

.   ..   .

Man's effort to dominate and to exploit the higher animals for his own benefit has brought much suffering to his fellow creatures. Those who have striven to relieve the animals have looked everywhere for substitutes and, as this book testifies, have found not a few, starting with humans themselves and running the gamut of technology. But help now comes from a most unexpected quarter - from spliced genes, pathogenic bacteria and malignant cancer cells. Thanks, ingenious scientists, for these humane and truly elegant experiments! Here at last are investigations which one can read about with admiration rather than despair. They are recorded with enthusiasm - and with new hope for the animals.

Adams, R. (1976). *Watership Down*. New York; Avon.

Ainsworth, M. (1976). "Discussion of papers by Suomi and Bowlby." In: *Animal Models in Human Psychobiology*. (G. Serban and A. Kling, eds.) New York; Plenum Press. p.37

Alexander, F. (1948). *Fundamentals of Psychoanalysis*. New York; Norton.

Altman, L. (1970). Twelve dogs develop lung cancer in group of 86 taught to smoke. *NY Times*, Feb. 6.

Altura, B. (1976). Sex and estrogens in protection against circulatory stress reactions. *Am. J. Physiol.* 231:842

American Cancer Society. (1978). *Cancer Facts and Figures, 1979*. New York; ACS.

American Museum of Natural History. (1977). *Annual Report, July 1976-June 1977*. New York; AMNH.

Anderson, K., *et al.* (1976). Behavioral evidence showing the predominance of diffuse pain stimuli over discrete stimuli in influencing perception. *J. Neurol. Res.* 2:283

Anderson, N. G. and N. L. (1979). Measurement instead of 'mouse data' on human mutation. Letter to *NY Times*, Apr. 27.

Animal Welfare Institute. (1973). *Analysis of the Annual Reports of Registered Research Institutions*. Washington; AWI.

Animal Welfare Institute. (1979). *Personal Communication* to author.

Anon. (1973). Human toxicology: research versus testing. *Lancet*, 1(7808):871.

Anon. (1974). India cuts Rhesus exports. *Natl. Soc. Med. Res. Bull.* 25(4):1.

Anon. (1975). Annual animal use report due on revised USDA forms. *Natl. Soc. Med. Res. Bull.* 26(1):1.

Anon. (1977a). Animal resources program. *Research Resources Reporter,* 1(4):1.

Anon. (1977b). Monkeys get radiation in neutron bomb tests. *Washington Post*, June 22.

Anon. (1978a). First of two CLINSPEC resources open. *Research Resources Reporter.* 2(4):8.

Anon. (1978b). FRAME Symposium. *ATLA Abstracts.* 6(1):10.

Anon. (1978c). Researchers say new drug kills millions of cancer cells in mice. *NY Times,* Apr.3, A17.

Anon. (1978d). Trends in primate imports into the United States. *Laboratory Primate Newsletter,* 17(3):10.

Anon. (1979a). Blood defends military radiation experiments of monkeys. *Newsletter Int. Primate Protection League,* 6(2):4.

Anon. (1979b). Genetic code-breakers crack hepatitis B. *New Scientist*, 83(1170):655.

Anon. (1979c). Lab animals. *Bull. ISAP.* 1(5):3.

Anon. (1979d). Macaque crisis, real or phony? *Newsletter Int. Primate Protection League,* 6(2):5.

To identify organizations abbreviated here, see Index.

Anon. (1979e). Workers' blood monitors chemical danger. *New Scientist*, 82(1151):185.

Anon. (1980a). Animals are suffering; HSUS seeks to end rabbit blinding tests. *The HSUS Close-up Report*, March.

Anon. (1980b). Hepatitis B vaccine from tumour cells. *New Scientist*, 85(1193):397.

Anon. (1980c). Monkeys killed with weed killer. *Newsletter Int. Primate Protection League.* 7(1):5.

Anon. (1980d). Mono vaccine on its way? *Science Digest Special*. Spring:98.

Anon. (1980e). New rabies vaccine awaits FDA action. *Natl. Soc. Med. Res. Bull.* 31(4):4.

Anon. (1980f). Psychologist criticises cruelty to primates at Brooks Air Force Base. *Newsletter Int. Primate Protection League,* 7(1):6.

Anon. (1980g). Scale up for new, old gene-splice products. *Sci. News,* 117:165.

Anon. (1980h). Whale imports blow from synthetic wax. *New Scientist,* 85(1199):929.

Aronson, L., *et al.* (1974a). *Behavioral effects of selected denervation.* (Grant application to USPHS). Communication from USPHS; Washington.

Aronson, L. and Cooper, M. (1974b). Olfactory deprivation and mating behavior in sexually experienced male cats. *Behav. Biol.* 11:459.

Assn. of Food and Drug Officials of the U.S. (1979). *Appraisal of the Safety of Chemicals in Foods, Drugs and Cosmetics.* Topeka, Kansas; AFDO.

Avon Products. (1973). *Agreement for Approval for Use of Living Animals in Scientific Tests.* Albany, NY; State Dept. of Health.

Baker, J. (1948). *The Scientific Basis of Kindness to Animals.* London; UFAW.

Baker, J., *et al.* (1949). Letters to the Editor: Experiments on animals. *Lancet,* 2(6571):259.

Bartlett, J. and Doty, R. (1974). Influence of mesencephalic stimulation on unit activity in striate cortex of squirrel monkeys. *J. Neurophysiol.* 37:642.

Barnes, C. and Eltherington, L. (1973). *Drug Dosage in Laboratory Animals.* Berkeley: University of California Press.

Bayard, S. and Hehir, R. (1976). Evaluation of proposed changes in the modified Draize Rabbit Irritation Test. *Toxicol. Appl. Pharmacol.* 37:186.

Bayly, M. (1952). *Experimental Shock in Animals: the Noble-Collip Drum.* London; Natl. Anti-Vivisection Soc.

Beckman Instruments. (1979). *Beckman Microtox Acute Water Toxicity Monitor.* Carlsbad, Cal.

Beddell, C.,*et al.* (1976). Compounds designed to fit a site of known structure in human haemoglobin. *Br. J. Pharmacol.* 57:201.

Bercaw, J.,*et al.* (1977). "Estimating injury from burning garments and development of concepts for flammability tests." In: *Fire Standards and Safety*. (A. F. Robertson, ed.) American Society for Testing and Materials. p.55.

Beston, H. (1976). *The Outermost House*. New York; Penguin.

Blanchard, R. and C. (1977). Aggressive behavior in the rat. *Behav. Biol.* 21:197.

Bourne, G., ed. (1977). *Progress in Ape Research*. New York; Academic Press.

Boyd, D. (1974). *Rolling Thunder*. New York; Random House.

Boyd, S.,*et al.* (1977). Use of cold-restraint to examine psychological factors in gastric ulceration. *Physiol. Behav.* 18:865.

Braun, A. and Nichinson, B. (1979). A new *in vitro* assay for teratogens. (Abstract). *Teratology*, 19:20A.

Brown, R. (1977). *Personal Communication* to Christine Stevens.

Brown, W.,*et al.* (1976). Protein metabolism in burned rats. *Am. J. Physiol.* 231:476.

BRS/LVE Division of Tech Serv., Inc. (1976). *Animal Environment*. (Catalog No. 1.)

Bruner, A.,*et al.* (1975). Delayed match-to-sample early performance decrement in monkeys after ^{60}Co irradiation. *Rad. Res.* 63:83.

Bruner, A. (1977). Immediate dose-rate of ^{60}Co on performance and blood pressure in monkeys. *Rad. Res.* 70:378.

Butler, C.,*et al.* (1977). Morphologic aspects of experimental esophageal lye strictures. II. Effect of steroid hormones.... *Surgery*, 81:431.

California, University of, and American Society of Microbiology. (1960). *ANAPHYLAXIS IN GUINEA PIGS*. 7 min. sound. color. 16 mm. (film).

Canadian Council on Animal Care. (1978). *Ethics of Animal Experimentation*. Ottawa; CCAC.

Cataldo, M., *et al.* (1978). Biofeedback teaches muscle control. *Research Resources Reporter*, 2(12):9.

Chance, M. (1956). Environmental factors influencing gonadotrophin assay in the rat. *Nature*. 177:455.

Chi, C., *et al.* (1976). Neuroanatomic projections related to biting attack elicited from ventral midbrain in cats. *Brain Behav. Evol.* 13:91.

Clayton, R. (1976). "A new rapid *in vitro* assay system for teratogenic compounds:" In: *Tests of Teratogenicity in Vitro*. (J. D. Ebert and M. Marois, eds.) Amsterdam; North Holland. p.473.

Cochin, J. (1973). "The use of the animal model in assessing analgesic potency and dependence liability." In: *Research Animals in Medicine*. (L. Harmison, ed.) Washington; GPO. p.701.

Coid, C. (1978). Symposium: Tests in laboratory animals - are they valid for man? *J. R. Soc. Med.* 71:675.

Collins, J. (1980). USFDA, Drug Bioanalysis Branch. *Personal Communication* to author.

Committee for the Reform of Animal Experimentation. (1977). *The LD 50 Test*. London; HMSO.

Coulston, F. and Serrone, D. (1969). The comparative approach to the role of nonhuman primates in evaluation of drug toxicity in man - a review. *Ann. NY Acad. Sci.* 162:681.

Croft, P. (1964). *An Introduction to the Anesthesia of Laboratory Animals*. London; UFAW.

Delgado, J. (1966). Aggressive behavior evoked by radio stimulation in monkey colonies. *Am. Zool.* 6:669.

Demling, R., *et al.* (1979). Effect of heparin on edema after second- and third-degree burns. *J. Surg. Res.* 26:27.

Devor, M. and Murphy, M. (1973). The effect of peripheral olfactory blockade on the social behavior of the male golden hamster. *Behav. Biol.* 9:31.

Dillingham, E., *et al.* (1973). Toxicity of methyl- and halogen-substituted alcohols in tissue culture.... *J. Pharm. Sci.* 62:22.

Diner, J. (1979). *Physical and Mental Suffering of Experimental Animals*. Washington; Animal Welfare Inst.

Donahue, E., *et al.* (1978). Detection of mutagenic impurities in carcinogens and noncarcinogens by high-pressure liquid chromatography and the *Salmonella/*Microsome test. *Cancer Res.* 38:431.

Dorworth, T. and Overmier, J. (1977). On "learned helplessness": The therapeutic effects of electroconvulsive shocks. *Physiol. Psychol.* 5:355.

Doty, R. (1975). Use of curariform agents. *Exp. Neurol.* 47:i.

Douglas, J., *et al.* (1977). Airway responses of the guinea pig *in vivo* and *in vitro*. *J. Pharmacol. Exp. Ther.* 202:116.

Drazen, J. and Schneider, M. (1978). Comparative responses of tracheal spirals and parenchymal strips to histamine and carbachol *in vitro*. *J. Clin. Invest.* 61:1441.

Drotman, R. (1977). "Metabolism of cutaneously applied surfactants." In: *Cutaneous Toxicity*. (V. A. Drill and P. Lazar, eds.) New York; Academic Press. p.95.

Duvall, D., *et al.* (1978).
Androgen and concurrent
androgen-progesterone maintenance of attack-eliciting
characteristics in male
mouse urine. *Behav. Biol.*
22:343.

Eagleton, T. (1977). "Cosmetic
legislation: a Congressional
viewpoint." In: *Cutaneous
Toxicity.* (V. A. Drill and
P. Lazar, eds.) New York;
Academic Press. p.265.

Eckhardt, E., *et al.* (1958).
Etiology of chemically induced writhing in mouse and
rat. *Proc. Soc. Exp. Biol.
Med.* 98:186.

Edelson, E. (1978). The human
animal and cancer tests. *NY
Daily News,* July 2.

Elder, R. (1979). CIR: an experiment in industrial self
regulation. *Cosmetic Technology.* 1(3):20.

Evarts, E. (1973). Brain mechanisms in movement. *Sci. Am.*
229:96.

Fentress, J. (1973). Development of grooming in mice with
amputated forelimbs.
Science, 179:704.

Flaxman, B. (1972). Replication
and differentiation *in vitro*
of epidermal cells from
normal human skin and from
benign (psoriasis) and malignant (basal cell cancer)
hyperplasia. *In Vitro.*
8:237.

Fortuna, M. (1977). Elicitation
of aggression by food deprivation in olfactory bulbecto-
mized male mice. *Physiol.
Psych.* 5:327.

Fox, M., *et al.* (1979). *Evaluation of Awarded Grant Applications Involving Animal
Experimentation.* Washington;
ISAP.

Frank, O. (1980). *The Use of
Protozoa for Screening of
Drugs and Metabolites.*
Address...Jan.19. AFAAR.

Freeman, A., *et al.* (1977).
Heteroploid conversion of
human skin cells by methyl-
cholanthrene. *Proc. Natl.
Acad. Sci.* 74:2451.

Freud, S. (1926). *Inhibitions,
Symptoms and Anxiety.*
London; Hogarth Press.

Frosch, P. and Kligman, A.
(1977). "The chamber-scari-
fication test for assessing
irritancy of topically applied substances." In:
Cutaneous Toxicity. (V. A.
Drill and P. Lazar, eds.)
New York; Academic Press.
p.127.

Gailer, K., *et al.* (1971). The
relative potency of some
beta-adrenoceptor agonists on
isolated human vein. *Eur. J.
Pharmacol.* 16:136.

Gehring, P., *et al.* (1973).
Toxicology: cost/time. *Food
Cosmet. Toxicol.* 11:1097.

Gilbert, W. and Villa-Komaroff,
L. (1980). Useful proteins
from recombinant bacteria.
Scientific American,
242(4):74.

Giovacchini, R. (1969). Premar-
ket testing procedures of a
cosmetic manufacturer.
Toxicol. Appl. Pharmacol.
Suppl. No. 3:13.

Giovacchini, R. (1972). Old and
    new issues in the safety
    evaluation of cosmetics and
    toiletries. *CRC Crit. Rev.
    Toxicol.* 1:361.

Glass, D. (1977). *Behavior
    Patterns, Stress and
    Coronary Disease*. Hills-
    dale,N.J.; Lawrence Erlbaum.

Glassman, R., *et al*. (1969). A
    safe and reliable method for
    temporary restraint of
    monkeys. *Physiol. Behav.*
    4:431.

Goltz, F. (1892). Der hund
    ohne Grosshirn. *Pfleugers
    Arch.* 51:570.

Grice, H. (1975). *The Testing
    of Chemicals for Carcinogen-
    icity, Mutagenicity and
    Teratogenicity*. Ottawa,
    Canada: Dept. of Health and
    Welfare.

Griffith, J. and Buehler, E.
    (1977). "Prediction of skin
    irritancy and sensitizing
    potential by testing with
    animals and man." In:
    Cutaneous Toxicity. (V. A.
    Drill and P. Lazar, eds.)
    New York; Academic Press.
    p.155.

Guess, W., *et al*. (1965). Agar
    diffusion method for toxicity
    screening of plastics on
    cultured cell monolayers.
    J. Pharm. Sci. 54:1545.

Harrison, B. (1971). Obtaining
    primates for research. *WFPA
    News*, No. 12:41.

Harriss, C., *et al*. (1978).
    Carcinogenesis studies in
    human cells and tissues.
    Cancer Res. 38:474.

Harrison, R. and M. (1978).
    Computer simulation as an
    aid to the replacement of
    experimentation on animals
    and humans. *ATLA Abstracts*.
    6(2):22.

Hayes, H. (1977). The pursuit
    of reason. *NY Times Mag.*
    June 12, p.21.

Hayflick, L. and Moorhead, P.
    (1961). The serial cultiva-
    tion of human diploid cell
    strains. *Exp. Cell Res.*
    25:585.

Hayflick, L. (1970). The
    choice of the cell sub-
    strate for human virus vac-
    cine production. *Lab.
    Practice*. 19(1):58.

Hazleton Laboratories Europe
    Ltd. (1976). *Personal Com-
    munication* to Dr. Hadwen
    Trust for Humane Pesearch.

Hazleton Laboratories Europe
    Ltd. (1977). *Personal Com-
    munication* to Dr. Hadwen
    Trust for Humane Research.

Heideman, M. (1979). The effect
    of thermal injury on hemo-
    dynamic, respiratory, and
    hematologic variables in
    relation to complement acti-
    vation. *J. Trauma*. 19:239.

Heim, A. (1978). *The Proper
    Study of Psychology*. London;
    British Association for the
    Advancement of Science.

Herman, P. (1979). UK Home
    Office. *Personal Communica-
    tion* to author.

Hillman, H. (1970). *Scientific
    Undesirability of Painful
    Experiments*. Zurich: WFPA.

Hinde, R. (1976). "The use of differences and similarities in comparative psychopathology." In: *Animal Models in Human Psychobiology*. (G. Serban and A. Kling, eds.) New York; Plenum Press. p.187.

Houser, V., *et al.* (1976). Effects of chlorpromazine on fear-motivated behavior, urinary cortisol, urinary volume and heart rate in the dog. *Psychol. Rep.* 38:299.

Houser, V. (1978). "The effects of drugs on behavior controlled by aversive stimuli." In: *Contemporary Research in Behavioral Pharmacology*. (D. E. Blackman and D. J. Sanger, eds.) New York; Plenum Press.

Howard, W., *et al.* (1971). Primate restraint system for studies of metabolic responses during recumbency. *Lab. Anim. Sci.* 21(1):112.

Hruza, Z. and Zweifach, B. (1970). Catecholamines and dibenzyline in trauma and adaptation to trauma. *J. Trauma*, 10:412.

Huang, Y., *et al.* (1975). *In vivo* location and destruction of the *locus coeruleus* in the stumptail macaque. *Brain Res.* 100:157.

Hughes, T. (1976). *The Use of Animals in Scientific Research in Canada*. Ottawa; Animal Welfare Foundation of Canada.

Humphrey, J. (1978). "Some challenges for practical immunologists." In: *Dimensions in Health Research*. (H. Weissbach and R. M. Kunz, eds.) New York; Academic Press. p.109.

Hynan, M. (1976). The influence of the victim on shock-induced aggression in rats. *J. Exp. Anal. Behav.* 25:401.

Inman, W. (1971). Why report adverse drug reactions? *Adverse Drug React. Bull.* 27:76.

Institute of Laboratory Animal Resources. (1975). *Nonhuman Primates*. Washington; NAS.

Institute of Laboratory Animal Resources (1978). *Guide for the Care and Use of Laboratory Animals*. Washington; GPO.

Interagency Primate Steering Committee. (1977). *National Primate Plan*. Washington; NIH.

Interagency Regulatory Liaison Group. (1979). *Scientific Bases for Identifying Potential Carcinogens and Estimating Their Risks*. (mimeo.)

Ishii, D. and Corbascio, A. (1971). Some metabolic effects of halothane on mammalian tissue culture cells *in vitro*. *Anesthesiology*, 34:427.

Iversen, S. and L. (1975). *Behavioral Pharmacology*. New York; Oxford.

Iversen, S. (1976). "Stress and behavior." In: *The Welfare of Laboratory Animals*. London; UFAW. p.64.

Jaggard, R., *et al.* (1950). Clinical evaluation of analgesic drugs: a comparison of Nu-2206 and morphine sulphate administered to postoperative patients. *Arch. Surg.* 61:1073.

Kaplan, H. (1969). Anesthesia in invertebrates. *Fed. Proc.* 28:1557.

Kaplan, J. and Saba, T. (1976). Humoral deficiency and reticuloendothelial depression after traumatic shock. *Am. J. Physiol.* 230:7.

Kardiner, A. (1941). *The Traumatic Neuroses of War.* New York; Hoeber.

Karkinen-Jääskeläinen, M. and Saxén, L. (1976). "Advantages of organ culture techniques in teratology." In: *Tests of Teratogenicity in Vitro.* (J. D. Ebert and M. Marois, eds.) Amsterdam; North Holland. p.275.

Katz, R. (1976). Role of the mystacial vibrissae in the control of isolation induced aggression in the mouse. *Behav. Biol.* 17:399.

Keele, C. and Smith, R. (1962). *The Assessment of Pain in Man and Animals.* London; Livingston.

Kestenbaum, R., *et al.* (1973). Inference of refractory period, temporal summation, and adaptation from behavior in chronic implants. *J. Comp. Physiol. Psychol.* 83:412.

Kim, S. and Pleasure, D. (1978). Tissue culture analysis of neurogenesis. *Brain Res.* 145:15.

Kochhar, D. (1976). "Elucidation of mechanisms underlying experimental mammalian teratogenesis through a combination of whole embryo, organ culture, and cell culture methods." In: *Tests of Teratogenicity in Vitro.* (J. D. Ebert and M. Marois, eds.) Amsterdam; North Holland. p.485.

Köhler, G. and Milstein, C. (1975). Continuous cultures of fused cells secreting antibody of predefined specificity. *Nature.* 256:495.

Koprowski, H., *et al.* (1978). "Anti-viral and anti-tumor antibodies produced by somatic cell hybrids." In: Lymphocyte Hibridomas. (F. Melchers, *et al.*, eds.) New York; Springer-Verlag. *(Current topics in microbiology and immunology:* v. 81) p.17.

Kosterlitz, H., *et al.* (1975). Narcotic...potencies... measured by the guinea pig ileum and mouse vas deferens methods. *J. Pharm. Pharmacol.* 27(2):73.

Krupp, J. (1976). Nine-year mortality experience in proton-exposed *Macaca mulatta. Rad. Res.* 67:244.

Laity, J. (1971). A comparison of the effects of isoprenaline, WG253 and salbutamol on the tension and rate of rabbit isolated atria. *J. Pharm. Pharmacol.* 23:633.

Lansdown, A. (1972). An appraisal of methods for detecting primary skin irritants. *J. Soc. Cosmetic Chemists,* 23:739.

Lash, J. and Saxén, L. (1971). Effect of thalidomide on human embryonic tissues. *Nature*. 232:634.

Lausch, E. (1972). *Manipulation, Dangers and Benefits of Brain Research*. New York; Viking.

Leakey, R. and Lewin, R. (1978). Origins of the mind. *Psychol. Today*. 12(2):49.

LeBlanc, J., *et al.* (1972). Beta-receptor sensitization by repeated injections of Isoproterenol and by cold adaptation. *Am. J. Physiol.* 222:1043.

LeCornu, A. and Rowan, A. (1978). *Historical Perspectives on the Use of Laboratory Animals and Their Tissues in the Development of Vaccines for Poliomyelitis*. London: FRAME.

Leong, S., *et al.* (1978). Detection of human melanoma antigens in cell-free supernatants. *J. Surg. Res.* 24:245.

Leuchtenberger, C., *et al.* (1973). Effects of marijuana and tobacco smoke on human lung physiology. *Nature*. 241:137.

LeVay, S. (1977). Effects of visual deprivation on polyribosome aggregation in visual cortex of the cat. *Brain Res*. 119:73.

Lilly, J. (1978). *Communication between Man and Dolphin*. New York; Crown.

Linden, E. (1974). *Apes, Men, and Language*. New York; Dutton.

Loan, W. and Dundee, J. (1967). The clinical assessment of pain. *Practitioner*. 198:759.

Lockard, V. and Kennedy, R. (1976). Alterations in rabbit alveolar macrophages as a result of traumatic shock. *Lab. Invest*. 35:501.

Lockley, R. (1965). *The Private Life of the Rabbit*. London; Andre Deutsch.

Loercher, D. (1972). White rabbits - blue eyes. *Christian Science Monitor*, June 12.

Loop, M. and Sherman, S. (1977). Visual discriminations during eyelid closure in the cat. *Brain Res*. 128:329.

Lore, R. and Luciano, D. (1977). Attack stress induces gastrointestinal pathology in domesticated rats. *Physiol. Behav*. 18:743.

Lumb, W. and Jones, E. (1973). *Veterinary Anesthesia*. Philadelphia; Lea & Febiger.

Madden, J., *et al.* (1973). Experimental esophageal lye burns. *Ann. Surg*. 178:277.

Magrath, D. (1978). *Personal Communication* to A. LeCornu and A. Rowan.

Maloney, J., *et al.* (1963). *Analysis of Chemical Constituents of Blood by Digital Computer*. Santa Monica, Cal.; Rand Corp.

Manson, J. and Simons, R. (1979). *In vitro* metabolism of cyclophosphamide in limb bud culture. *Teratology*. 19:149.

Marlin, G. and Turner, P. (1975). Comparison of the $beta_2$-adrenoceptor selectivity of rimiterol, salbutamol and isoprenaline by the intravenous route in man. *Br. J. Clin. Pract.* 2:41.

Marzulli, F. and Ruggles, D. (1973). Rabbit eye irritation test: Collaborative study. *JOAC.* 56(4):905.

Mason, J. (1972). Corticosteroid response to chair restraint in the monkey. *Am. J. Physiol.* 222:1291.

Mass, M. and Lane, B. (1976). Effect of chromates on ciliated cells of rat tracheal epithelium. *Arch. Environ. Health.* 31:96.

Matthews, B. and Searle, B. (1976). Electrical stimulation of teeth. *Pain.* 2:245.

McCann, J. and Ames, B. (1976). A simple method for detecting environmental carcinogens as mutagens. *Ann. NY Acad. Sci.* 271:5.

McGreal, S. (1978). Monkeying with the rhesus. *Illus. Wkly. India.* 99(41):20.

McGreal, S. (1980). *Personal Communication* to author.

Melchers, F., *et al.* (1978). "Lymphocyte Hybridomas." New York; Springer-Verlag. *(Current topics in microbiology and immunology;* v. 81) p.XVII.

Mineka, S. and Suomi, S. (1977). An opponent-process interpretation of multiple peer separations in Rhesus monkeys. *Bull. Psychonomic Soc.* 10(4):254.

Mitruka, B., *et al.* (1976). *Animals for Medical Research.* New York; Wiley.

Moorcroft, W., *et al.* (1971). Ontogeny of starvation-induced behavioral arousal in the rat. *J. Comp. Physiol. Psychol.* 75:59.

Moore-Ede, M. and Herd, J. (1977). Renal electrolyte circadian rhythms: independence from feeding and activity patterns. *Am. J. Physiol.* 232:F128.

Morgan, Elizabeth, M.D. (1980). *The Making of a Woman Surgeon.* New York; Putnam.

Morgan, K. (1979). How dangerous is low-level radiation? *New Scientist.* 82(1149):18.

Morrison, J., *et al.* (1968). "The purposes and value of LD 50 determinations." In: *Modern Trends in Toxicology,* v. I. (E. Boyland and R. Goulding, eds.) London; Butterworth. p.1.

Murphy, M. (1976). Blinding increases territorial aggression in male Syrian golden hamsters. *Behav. Biol.* 17:139.

Murphy, M. and Schneider, G. (1970). Olfactory bulb removal eliminates mating behavior in the male golden hamster. *Science.* 167:102.

Nardone, R. (1977). "Toxicity testing *in vitro.*" In: *Growth, Nutrition and Metabolism of Cells in Culture.* (G. M. Martin and C. E. Ogburn, eds.) New York; Academic Press.

Nashold, B., *et al.* (1969).
Sensations evoked by stimulation in the midbrain of
man. *J. Neurosurg.* 30:14.

NAS Committee on Medical and
Biological Effects of Atmo-
spheric and Environmental
Pollutants. (1974). *Chromium.*
Washington; NAS. p.42.

National Cancer Advisory Board.
Subcommittee on Environmental
Carcinogenesis. (1976).
*General Criteria for
Assessing the Evidence for
Carcinogenicity of Chemical
Substances.* (Unpublished
report).

NRC. (1971). *A Guide to Envi-
ronmental Research on
Animals.* Washington; Natl.
Academy of Sciences.

NRC. (1977). *Principles and
Procedures for Evaluating
the Toxicity of Household
Substances.* Washington;
Natl. Academy of Sciences.

Noble, H., *et al.* (1977). Dermal
ischemia in the burn wound.
J. Surg. Res. 23:117.

Noguchi, P., *et al.* (1978).
Chick embryonic skin as a
rapid organ culture assay
for cellular neoplasia.
Science. 199:980.

Nolen, W. (1970). *The Making of
a Surgeon.* New York; Random
House.

PDR (Physicians' Desk Reference).
(1975). Oradell, New Jersey;
Medical Economics Co.

PDR (Physicians' Desk Reference).
(1978). Oradell, New Jersey;
Medical Economics Co.

Penfold, T. (1974). *Personal
Communication* of Dec. 10 to
Mayor and Council of Everett,
Washington.

Petricciani, J. (1978). *Personal Communication* to
author.

Pew, T. (1979). Biofeedback
seeks new medical uses for
concept of yoga. *Smith-
sonian.* 10(9):106.

Pieper, W. (1977). "Acute
effects of stimulants and
depressants on sequential
learning in great apes." In:
Progress in Ape Research.
(G. H. Bourne, ed.) New
York; Academic Press. p.167.

Poggio, C. and Mountcastle, V.
(1960). A study of the
functional contributions of
the lemniscal and spinothal-
amic systems....*Bull.
Hopkins Hosp.* 106:266.

Pratt, D. (1976). *Painful Experiments on Animals.* New
York; Argus Archives.

Provost, P. and Hilleman, M.
(1979). Propagation of human
hepatitis A virus in cell
culture *in vitro. Proc. Soc.
Exp. Biol. Med.* 160:213.

Rackow, H. (1979). *Personal
Communication* to author.

Rajan, T., *et al.* (1972). The
response of human pleura in
organ culture to asbestos.
Nature. 238:346.

Richards, W. (1975). *Can Theo-
retical Understanding Reduce
the Need for Experiments?*
(Address given at the General
Assembly of the IAAPEA,
Amsterdam, July, 24).

Richards, W. (1977). "Calcula-
tion of essential drug confor-
mations and electron distribu-
tions." In: *The Lord Dowding
Fund Conference Papers* Presented Jan. 24, Sect. C.

Richter, C. (1957). On the phenomenon of sudden death in animals and man. *Psychosom. Med.* 19:191.

Richter, C. (1971). Inborn nature of the rat's 24-hour clock. *J. Comp. Physiol. Psychol.* 75;1.

Richter, C. (1978). Evidence for existence of a yearly clock in surgically and self-blinded chipmunks. *Proc. Natl. Acad. Sci.* 75:3517.

Rofe, P. (1971). Tissue culture and toxicology. *Food Cosmet. Toxicol.* 9:683.

Rosenberg, K. (1974). Effects of pre- and postpubertal castration and testosterone on pup killing behavior in the male rat. *Physiol. Behav.* 13:159.

Roth, N. (1979). Hazleton Laboratories Corporation. *Paine Webber Mitchell Hutchins, Regional Research, Basic Analysis,* July 11, p.1.

Rowan, A. (1979a). Alternatives to Laboratory Animals. Washington; ISAP.

Rowan, A. (1979b). Beauty and the beasts. *Humane Soc. News.* 24(2):4.

Rubin, N. (1974). *Personal Communication* to author.

Ruesch, H. (1978). *Slaughter of the Innocent.* New York; Bantam.

Russell, J. (1978). Monkeying around. *Courier-Journal,* June 1.

Russell, W. and Burch, R. (1959). *The Principles of Humane Experimental Technique.* London; Methuen.

Ryder, R. (1975). *Victims of Science.* London; Davis-Poynter.

Sakakura, H. and Doty, R., Sr. (1976). EEG of striate cortex in blind monkeys; effects of eye movements and sleep. *Arch. Ital. Biol.* 114:23.

Sako, F., *et al.* (1977). Effects of food dyes on Paramecium caudatum. *Toxicol. Appl. Pharmacol.* 39(1):111.

Salisbury, D. (1978). Research with animals. *Christian Science Monitor,* Mar. 10.

Sarfeh, I. and Balint, J. (1977). Hepatic dysfunction following trauma: Experimental studies. *J. Surg. Res.* 22:370.

Schaller, G. (1973). *Golden Shadows, Flying Hooves.* New York; Knopf.

Schlumpf, M., *et al.* (1977). Explant cultures of catecholamine-containing neurons from rat brain. *Proc. Natl. Acad. Sci.* 74:4471.

Schmeck, H. (1979). Researchers seek a vaccine 'booster.' *NY Times,* Feb. 27, p.C1.

Schwindaman, D., *et al.* (1973). "The use of implantable telemetry systems for animal monitoring." In: *Research Animals in Medicine.* (L. Harmison, ed.) Washington, GPO. p.1271.

Scobie, S. (1972). Interaction of an aversive Pavlovian conditional stimulus with aversively and appetively motivated operants in rats. *J. Comp. Physiol. Psychol.* 79(2):171.

Scott, R. and Stewart, J.
    (1979). *Films for Humane
    Education.* New York;
    Argus Archives.

Seigler, H. (1977). "Immunology
    and melanoma." In: *Progress
    in Ape Research.* (G. H.
    Bourne, ed.). New York;
    Academic Press, p.227.

Seligman, M. and Beagley, G.
    (1975). Learned helplessness
    in the rat. *J. Comp.
    Physiol. Psychol.* 88:534.

Seligman, M., *et al.* (1975).
    Learned helplessness in the
    rat: time course, immuniza-
    tion, and reversibility. *J.
    Comp. Physiol. Psychol.*
    88:542.

Severo, R. (1980). Dispute
    arises over Dow studies on
    genetic damage in workers.
    NY Times, Feb. 5.

Sincock, A. and Seabright, M.
    (1975). Induction of chromo-
    some changes in Chinese ham-
    ster cells by exposure to
    asbestos fibres. *Nature.*
    257:56.

Slauson, D., *et al.* (1976).
    Inflammatory sequences in
    acute pulmonary radiation
    injury. *Am. J. Pathol.*
    82:549.

Smith, R. (1977). The monkey
    business. *The Sciences.*
    17(4):15.

Smyth, D. (1978). *Alternatives
    to Animal Experiments.*
    London; Scolar Press.

Sontag, J., *et al.* (1976).
    *Guidelines for Carcinogen
    Bioassay in Small Rodents.*
    Washington; GPO. DHEW No.
    (NIH) 76-801. NCI Tech. Rep.
    No. 1.

Spector, S. and Hull, E. (1972).
    Anosmia and mouse killing by
    rats. *J. Comp. Physiol.
    Psychol.* 80(2):354.

Staehelin, T. (1978). "Applica-
    tion of Recent Discoveries
    in Immunology." In: *Dimen-
    sions in Health Research.*
    (H. Weissbach and R. M. Kunz,
    eds.) New York; Academic
    Press. p.133.

Stilley, F. (1975). *The $100,000
    Rat.* New York; Putnam.

Stockton, W. (1980). On the brink
    of altering life. *NY Times
    Mag.,* Feb. 18, p.16.

Stoffer, G. and J. (1976). Stress
    and aversive behavior in non-
    human primates. *Primates.*
    17(4):547.

Stokman, C. and Glusman, M.
    (1969). Suppression of hypo-
    thalamically produced flight
    responses by punishment.
    Physiol. Behav. 4:523.

Stoner, H. (1961). Critical
    analysis of traumatic shock
    models. *Fed. Proc.* 20,
    Suppl. 9, p.38.

Styles, J. (1977). A method for
    detecting carcinogenic organic
    chemicals using mammalian
    cells in culture. *Br. J.
    Cancer* 36:558.

Suomi, S. (1974). "Factors
    affecting responses to social
    separation in rhesus monkeys."
    In: *Animal Models in Human
    Psychobiology.* (G. Serban and
    A. Kling, eds.) New York;
    Plenum Press. p.9.

Suomi, S., *et al.* (1976). Effects
    of maternal and peer separa-
    tions on young monkeys. *J.
    Child Psychol. Psychiatry,*
    17:101.

Swenson, R. and Randall, W. (1977). Grooming behavior in cats with pontile lesions and cats with tectal lesions. *J. Comp. Physiol. Psychol.* 91:313.

Tavris, C. (1973). "Harry, you are going to go down in history as the father of the cloth mother." *Psychol. Today,* 6(11):65.

Tayler, R. and Piper, D. (1977). The carcinogenic effects of cigarette smoke...on human gastric mucosal cells in organ culture. *Cancer.* 39:2520.

Taylor, D. (1977). "Royal College of Surgeons." In: *The Welfare of Laboratory Animals; Legal, Scientific and Humane Requirements.* Potters Bar, Hertfordshire; UFAW, p.103.

Thomas, L. (1974). *The Lives of the Cell.* New York; Viking.

Tint, H., *et al.* (1974). A new tissue culture (WI-38) rabies vaccine....*Symp. Ser. Immunbiol. Stand.,* 21:132.

Trager, W. and Jensen, J. (1976). Human malaria parasites in continuous culture. *Science.* 193:673.

Trier, J. (1976). Organ-culture methods in the study of gastrointestinal-mucosal function and development. *N. Eng. J. Med.* 295:150.

Tucker, A. (1977). Animal army in nuclear test. *Guardian,* May 18.

Tucker, J. (1976). The immense puzzle. *Yale Alumni Magazine,* 40(4):33.

Turinsky, J., *et al.* (1977). Dynamics of insulin secretion and resistance after burns. *J. Trauma,* 17:344.

Turner, L. and Soloman, R. (1962). Human traumatic avoidance learning. *Psychol. Monogr.* 76 (Whole No. 559):1.

Ulrich, R. (1966). Pain as a cause of aggression. *Am. Zoologist,* 6:643.

Ulrich, R. (1973). Toward experimental living. *Behav. Modification Monogr.* 2:1.

Ulrich, R. (1979a). *Personal Communication* to author.

Ulrich, R. (1979b). Some thoughts on human nature and its control: I am my neighbor and my neighbor is me. *J. Humanistic Psychol.* 19:29.

Underwood, H. and Menaker, M. (1971). Photoreception in sparrows. *Science.* 172:293.

United Action for Animals. (1973). *Abstracts Regarding Testing of Environmental Chemicals.* New York; UAA.

UK Home Office. (1974). *Personal Communication* to author.

UK Home Office. (1978). *Experiments on Living Animals; Statistics, 1977.* London; HMSO.

UK DHSS. (Dept. of Health and Social Security). (1972). *The Use of Fetuses and Fetal Material for Research.* London; HMSO.

UK DHSS. (1977). *Compendium of Licensing Requirements for the Manufacture of Biological Medicinal Products.* London; DHSS.

US Congress. House of Representatives. (1966). *Hearings:...Regulate the Transportation, Sale and Handling of Dogs and Cats Used for Research and Experimentation*. Washington; GPO.

USDA/APHIS (1972). *Annual Report of Research Facility under Animal Welfare Act*. Hyattsville, Md.
    Revlon Research Center

USDA/APHIS (1973). *Annual Report of Research Facility under Animal Welfare Act*. Hyattsville, Md.
    Avon Products
    Bristol Laboratories
    Johnson & Johnson
        Research Foundation
    Revlon Research Center
    Warner-Lambert Research
        Institute

USDA/APHIS (1976). *Annual Report of Research Facility under Animal Welfare Act*. Hyattsville, Md.
    Avon Products
    Battelle Memorial Institute
    Bayvet Corp.
    Bristol Laboratories
    Dow Chemical Co.
    Endocrine Laboratories
    Florida, University of
    Hill Top Research, Inc.
    Hoffmann-La Roche, Inc.
    Imperial Chemical
        Industries, US
    Iowa, University of
    Pennsylvania, University of
    Procter & Gamble Co.
    Revlon Research Center
    Robins (A.H.) Research
        Laboratories

USDA/APHIS (1976). (Cont.)
    Tulane University
    Warner-Lambert Research
        Institute

USFDA. (1978). Nonclinical laboratory studies; good laboratory practice regulations. *Federal Register*, 43(247): 59986.

US Laws, Statutes, etc. (1972). *Code of Federal Regulations*. Title 9 Animals and Animal Products, Chap. I Animal and Plant Health Inspection Service. Washington.

US Laws, Statutes, etc. (1978). *Code of Federal Regulations*. Title 21 Food and Drugs, Chap. I Food and Drug Administration, Sect. 170-299. Washington.

US Laws, Statutes, etc. (1978). *Code of Federal Regulations*. Title 21 Food and Drugs, Chap. I Food and Drug Administration, Sect. 740.10. Washington.

US Laws, Statutes, etc. (1979). *Code of Federal Regulations*. Title 9 Animals and Animal Products, Chap. I Animal and Plant Health Inspection Service, Sect. 2.28. Washington.

US Laws, Statutes, etc. (1979). *Code of Federal Regulations*. Title 9 Animals and Animal Products, Chap. I Animal and Plant Health Inspection Service, Sect. 113.99(c)4. Washington.

US Laws, Statutes, etc. (1979). *Code of Federal Regulations.* Title 16 Commercial Practices, Chap. II Consumer Product Safety Commission, Sect. 1500.40-41. Washington.

US Laws, Statutes, etc. (1979). *Code of Federal Regulations.* Title 16 Commercial Practices, Chap. II Consumer Product Safety Commission, Sect. 1500.42. Washington.

US NIH. (1979). *Research Grants, 1978.* Washington, GPO.

Universities Federation for Animal Welfare. (1963). *Evidence Tendered on Behalf of...to UK Parliament, Departmental Committee of Inquiry into Experiments on Animals.* London; UFAW.

van Lawick, H. and van Lawick-Goodall, J. (1970). *Innocent Killers.* London; Collins.

Von Noorden, G., *et al.* (1977). Effect of lid suture on retinal ganglion cells in *Macaca mulatta. Brain Res.* 122:437.

Walton, J. and Buckley, I. (1975). Cell models in the study of mechanisms of toxicity. *Agents Actions,* 5:69.

Webster, R. and Maibach, H. (1977). "Percutaneous absorption in man and animal: a perspective." In: *Cutaneous Toxicity.* (V. A. Drill and P. Lazar, eds.) New York; Academic Press. p.111.

Veil, C. and Scala, R. (1971). Study of intra- and interlaboratory variability in the results of rabbit eye and skin irritation tests. *Toxicol. Appl. Pharmacol.* 19:276.

Weisenberg, M., ed. (1975). *Pain; Clinical and Experimental Perspectives.* St. Louis; Mosby.

Weiss, B., *et al.* (1977). Movement disorders induced in monkeys by chronic haloperidol treatment. *Psychopharmacol.* 53:289.

Weiss, J. (1971). On stress pathology in rats. *J. Comp. Physiol. Psychol.* 77(1):14.

Weiss, J., *et al.* (1976). "Coping behavior and neurochemical changes." In: *Animal Models in Human Psychobiology.* (G. Serban and A. Kling, eds.) New York; Plenum Press, p.141).

Whitaker, A. (1977). "Application of cell culture methods to virology." In: *The Pharmaceutical Applications of Cell Techniques.* Potters Bar, Hertfordshire; UFAW. p.28.

Wilkinson, J. (1980). Engineering a genetic revolution. *New Scientist.* 85(1197):728.

Williams, J. and Little, J. (1977). Selective protection of cultured human cells from the toxic effects of ultra-violet light by proflavine pretreatment. *Radiat. Res.* 72:154.

Wilsnack, R., *et al.* (1973). Human cell culture toxicity testing of medical devices and correlation to animal tests. *Biomater. Med. Devices Artif. Organs.* 1:543.

Wolfe, R. and Miller, H. (1976). Cardiovascular and metabolic responses during burn shock in the guinea pig. *Am. J. Physiol.* 231:892.

Wolfe, R. and Burke, J. (1977a).
Effect of burn trauma on glucose turnover, oxidation and
recycling in guinea pigs.
Am. J. Physiol. 233:E80.

Wolfe, R., *et al.* (1977b). Glucose and lactate kinetics in
burn shock. *Am. J. Physiol.*
232:E415.

Yamasaki, E. and Ames, B.
(1977). Concentration of
mutagens from urine by adsorption with the nonpolar
resin XAD-2. *Proc. Natl.
Acad. Sci.* 74:3555.

---

[1] *In parentheses: experiments on intact animals for which alternatives are proposed.*

[1]
In parentheses: experiments on intact animals for which alternatives are proposed.

[1]*In parentheses: experiments on intact animals for which alternatives are proposed.*

[1] *In parentheses: experiments on intact animals for which alternatives are proposed.*